Windfall

Also by Robert W. Righter

National Parks in the American West (St. Louis, 1979)

Crucible for Conservation: The Creation of Grand Teton National Park (Boulder, Colo., 1982)

The Making of a Town: Wright, Wyoming (Boulder, Colo., 1985)

(ed.) *Teton Country Anthology* (Boulder, Colo., 1990)

Wind Energy in America: A History (Norman, Okla., 1996)

(coed. with Martin J. Pasqualetti and Paul Gipe) *Wind Power in View: Energy Landscapes in a Crowded World* (San Diego, 2002)

The Battle over Hetch Hetchy: America's Most Controversial Dam and the Birth of Modern Environmentalism (New York, 2005)

Windfall

Wind Energy in America Today

Robert W. Righter

UNIVERSITY OF OKLAHOMA PRESS : NORMAN

This book is published with the generous assistance of
The McCasland Foundation, Duncan, Oklahoma.

Library of Congress Cataloging-in-Publication Data

Righter, Robert W.
  Windfall : wind energy in America today / Robert W. Righter.
      p. cm.
  Includes bibliographical references and index.
  ISBN 978-0-8061-4192-3 (pbk. : alk. paper)
  1. Wind power—United States. 2. Wind turbines—Environmental aspects—
United States. 3. Electric power production—United States. I. Title.
  TJ820.R54 2011
  621.31'2136—dc22

                                2011000238

The paper in this book meets the guidelines for permanence and durability
of the Committee on Production Guidelines for Book Longevity of the
Council on Library Resources, Inc. ∞

For my brother
Richard Righter

Contents

Illustrations

Acknowledgments

I would like to start out not with acknowledgments but with a confession. Some of the facts and examples in this book are out of date even before publication. This is indeed frustrating for a historian accustomed to publishing relevant work that will hold up for a long time. However, the facts on the growth of wind energy shift almost as fast as the ink dries. A time exposure, say of ten years, is an impossible task, akin to trying to get an unobstructed photograph of a child in full play mode. Yet although the facts and figures are fleeting, I believe the ideas and concepts remain identifiable. For instance, when I recently explained a current wind energy controversy, a friend commented that it was "no different than Tehachapi (Pass) in 1983." Places, settings, companies, and people change, but most of the issues confronting wind energy are constant. So this is my warning. The specific information given reaches up to May 2009. That was my cut-off date, and I acknowledge that much has happened since then.

I investigated many of the book's current issues from Internet sources. Without that this work would not exist in its present form. Printed sources on contemporary wind energy issues are limited. I have tried to apply discretion in using Internet sources, well aware of complaints that such material is not necessarily complete or accurate and that opinions are certainly suspect. Furthermore, in the paperless world of the Internet sometimes the citations one provides in endnotes simply disappear into the cosmos, untraceable. One of the key websites I used has now confessed that many of its formerly posted items are "irretrievably gone from the site."

I have many anonymous people to thank. For the past decade I have asked innumerable strangers (and some friends) what they think of wind energy. It is rare, particularly in the West, that they do not have an opinion. Their views, one way or another, have been incorporated into the book. Friends such as Ken Down, Asa Barnes, Jr., Kenton and Marlee Clymer, and Dan and Marly Merrill have added to the mosaic of opinion represented. My brother, Richard Righter, provided an informal clipping service as well as interest that never faltered. I am also in debt to Mike Pasqualetti, Paul Gipe, and T. Lindsay Baker, who read the manuscript and offered comments, often critical. They improved the work immensely.

I owe special thanks to Professors James Hopkins and Kathleen Wellman of the Department of History at Southern Methodist University for travel grants that permitted me to do field research. Chuck Rankin, senior editor at the University of Oklahoma Press, deserves my gratitude for suggesting this follow-up to my earlier book on wind energy and then for providing the inspiration to get it done. Steven Baker and Sally Antrobus shepherded the manuscript into publication, for which I am grateful. Sherry Smith, my wife of twenty-five years, has sometimes been a demanding critic but has always been a loving, understanding partner.

Introduction

In twelfth-century England the nobility defined class distinctions. The clergy and the aristocracy occupied the upper echelons, and peasants the lower end. There was not much middle ground between these extremes. However, at least several times a year, nature intervened to bring their interests in harmony. When violent storms swept across the island, they wreaked havoc with the estates of the wealthy, strewing the manicured forests with broken limbs and uprooted trees. Peasants needed warmth in their huts that a wood fire might provide. By tradition after such storms, the nobility opened their wooded land to the peasants, who eagerly harvested the branches and tree trunks for firewood. From this practice comes the derivation of the word *windfall*.[1] In our era we speak of a windfall as an unexpected gain, usually in the form of money.

Many in the twenty-first century see the growth and development of wind energy as an unexpected gain. Since the beginning of time, the wind has been blowing, filling pressure vacuums and creating new ones. Although the English peasants may have looked on the wind as a gain, over history the wind has more often been viewed as an annoyance. Certainly the English ruling class saw a storm as an inconvenience that would require them to extend a privilege to the lower class, one they surely were reluctant to give. Thus on the surface, the greatest benefit of a windfall went to the poor. However, it was a win-win situation. In effect, peasants were pruning or harvesting the woodlots for the wealthy. Without cost, the

nobility once again had tidy estates, while landless peasants might anticipate a winter of warmth.

In a way, I feel somewhat like those twelfth-century English people, for
I am not an engineer. Nor am I a mathematician. As a student, I found getting through algebra and geometry a struggle, and that is where it ended.
This is just a warning. Creating electricity from the wind is a game of
mathematics and engineering, and a number of books explain that game
quite adequately. This is not one of them. I am a historian with a specialty
in the American West and environmental history. In my profession I believe the past has lessons, though not necessarily answers, for the present. Looking at the nation's past experience convinces me that the idea of
windfall does have relevance for the energy situation today. Wind farms
offer numerous gains for both individuals and corporations, but particularly for the environment. Aside from solar panels, there is no more benign
way to create the electricity we want and need. In this book I return often
to the gains and losses of wind energy, but for now I stand on the belief
that, like the windfalls of the twelfth century, the turbines spreading across
the American landscape represent a win-win situation. They are a positive
response to the eventual decline of the oil age and the nation's growing
concern with an overheated earth.

As a historian I thought I had finished my work in 1996 when I published *Wind Energy in America: A History*. In that year, energy from the
wind still represented a novel idea, misunderstood by most Americans. In
the past decade, however, the wind energy business has made astounding
progress, advances that few anticipated. A second edition seemed worthwhile. This became particularly evident when, in the fall of 2006, the state
of Texas surpassed the wind energy capacity of California, which until
then had long been the leader. When I was researching *Wind Energy in
America*, Texas had zero commercial wind energy capacity. Now not only
does the state lead the nation in production, but with deregulation, my
wife and I buy all our Dallas home's electrical power from Green Mountain Energy. Though not the major supplier, Green Mountain provides
consumers with electricity produced from the wind, augmented with a
small percentage of hydro power. Texas is not unique. States across the
country have created business atmospheres conducive to attracting the
great turbines. Throughout the West, wind turbines have been sprouting
up on the landscape like exotic weeds. These turbines, with their futuristic
appearance, give hope for a relatively risk-free way to continue our depen-

dence on electricity. Whether the turbines can fulfill such a difficult challenge is doubtful, but they are icons of hope nevertheless, ubiquitous both on the landscape and in the media, especially in film and advertisements. As one *New York Times* journalist put it, "The wind turbine is the 'it' item of summer 2008."[2] There is nothing to indicate this positive symbol will change soon.

Remarkably, wind energy is the fastest-growing form of renewable energy in the world. With so much activity in the industry, it became apparent I had to do more than update an old book. I needed to write more on the contemporary situation, much more. In 1990 a wind farm was rare—an object of curiosity that could be found only in California's mountain passes. Two decades later one cannot travel the American West without, sooner or later, encountering the huge turbines spinning nearby or on a more distant hill or mesa. Beyond the West the turbines have spread to such midwestern states as Iowa, Minnesota, Wisconsin, and Illinois. Even the more populated East Coast has turbines working and projects on the drawing board, many of them controversial. Wind electricity has become a nationwide phenomenon.

In a modest way this book commemorates a new age of renewable energy, and wind energy represents one of the chief agents of that change. Given its timing, what should be the objectives of this new book? First, I want to give an idea of how a wind generator works and how the electricity it produces ends up powering our air conditioners on a warm day. To do that, we have to look at wind turbine *production*. Then we examine *distribution*, essentially how electricity travels from its creation to people's homes. Production and distribution of wind energy—or for that matter any energy—is extremely complicated, requiring the knowledge of electrical engineers trained in such matters. This became abundantly clear recently when I attended Professor Andrew Swift's wind energy class at Texas Tech University. Much of his lecture and all of his diagrams and mathematical equations were incomprehensible to me. And yet how many of us understand how a microwave oven, a telephone, a computer, or even a calculator works? We simply accept them. Although I avoid the more technical aspects of production and distribution, I do explore in some detail the social, cultural, political, economic, and environmental ideas and concepts that will ultimately determine the long-term fate of this new technology. Public policy, after all, has more to do with the success or collapse of wind energy than with technological advances. At the moment, wind energy is

Turbines are assisting the nation's effort to combat global warming, air pollution, and the depletion of natural resources. They have also become icons of patriotism, valued not only for their environmental gains but also because new manufacturing plants offer American jobs. (Photo by Frank A. Oteri, NREL.)

on a remarkable growth trajectory, but just how far will the nation wish to go?

A second objective is to analyze the pros and cons of this wind energy surge. Who would have thought in the 1990s that giant wind turbine "farms" would be radiating across the country, creating a scourge for some, a blessing for others? Whether welcome or not, wind developments are coming on line and being planned faster than American, Japanese, and European manufacturers can produce the turbines. What is happening? Are these huge turbines a good thing or, as they were in the early 1980s, are they machines that generate more tax credits than usable electricity? There are plenty of friends of wind energy, but there are also determined enemies. Today, even the announcement of a possible project stirs controversy. Surprisingly, the most heated battle is not in the American West but in Massachusetts. There the Cape Wind project, a plan for placing 130 turbines in Nantucket Sound, has spawned a book, numerous articles, and the ire of some of the wealthiest and most powerful families in the nation.[3]

The most prevalent con lies in the tremendous size of the new turbines and the consequent visual impact on landscapes (or seascapes) we love. Recently on a commercial flight from Dallas to Los Angeles I spotted a wind farm from thirty thousand feet. Although bird death ("avian mortality") and generator and blade noise are notable issues, it is the industrializing of rural landscapes that has mobilized neighbors to protest. Many people find the turbines visually annoying, and on the economic level they are convinced that their property values have diminished (or will drop) as nearby turbines rise to the sky. Through examples, I aim to give a sense of the objections. Beyond the aesthetic argument, critics maintain that the turbines simply do not produce electricity as promised. Exaggeration is common in the wind energy business. So what is the promise versus the reality? Some of the statistics give good cause for skepticism. Opponents argue that without massive federal, state, and local tax credits and various subsidies, wind energy would not exist. I consider in some detail the assistance given this relatively new energy source.

On the pro side, the environmental arguments for cleaner and cooler air, once thought unimportant, now become compelling. Natural energy sources (wind, water, solar, tides, and so forth) promise to alleviate the anguish caused by our nonrenewable, polluting power sources. Chapter 9 compares wind energy to more traditional forms of electricity production.

I look briefly at fossil fuels (oil, coal, natural gas) and nuclear plants as base load producers; the electrical output of these fuels can be controlled. No responsible person would suggest that wind energy alone can fulfill the nation's electrical needs. It will always supplement traditional base load sources, particularly coal. Yet this in itself is an argument for wind energy, for any sensible energy policy will promote diversity—a mingling of kinetic and heat sources. Wind energy can play an important role as one of a cluster of energy sources. It has many benefits that I underscore. We know this planet would be better off without *any* conversion of resources to electrical energy, but that is both unrealistic and contrary to human nature.

The third objective is to speculate on the future of wind energy from a cultural perspective: that is, how will we react emotionally and psychologically to a transformed landscape? Wind energy has almost limitless potential, but how much of the natural American landscape will we sacrifice for our energy needs? I also want to examine the possibility of changes in landscape perception, particularly regarding the overlay of a built landscape atop a natural one. Do humans have the capacity to rethink what constitutes a pleasing landscape? Is the fabric of landscape change acceptable in rural settings, particularly to the younger generation? Essentially, can we learn to accept (and even enjoy) an industrialized rural landscape? Can we blend the artificial with the natural? Or is the vision of an attractive landscape genetically imprinted, impervious to modification? I, of course, do not think it is. A love of natural landscape, however, is deeply ingrained in American history and culture. Throughout our nation's past, thinkers and writers (and viewers) have struggled with this dichotomy between our love of nature and our commitment to technology. Reconciling the two has not been easy. We hope for a "middle landscape" where tools and nature unite harmoniously.[4] I tend to believe that as Americans think about wind turbines and what they accomplish, we will accept their presence. They do, after all, represent intermediate technology: that is, they are not powered by an internal combustion engine or any other form of man-made energy. They are, if you will, the big brother of the water-pumping windmill, an example of a much-revered intermediate technology that has become a cultural icon in the American West and elsewhere. In many ways the new, immense, modern wind turbines seem, at first sighting, foreign to our perceptions of what an agreeable landscape should be. But on reflection, the American landscape is every bit as bold and powerful as the turbines that may inhabit it. The nation has the luxury of space. If we think about the natural way these turbines perform and what they *do*, we may find at least an uneasy truce, if not perfect harmony.

Even if sensible people oppose wind turbines, the reality is that they may have little electricity choice. Although the United States has been reluctant to confront the realities of global warming, a new generation cannot accept inaction. In January 2009 a new administration took power in Washington, D.C., and even before assuming the presidency, Barack Obama signaled his intention to attack our polluting habits by doubling the nation's renewable energy production in the next three years. His staff announced that they hoped to add 20 gigawatts (20,000 megawatts) or more of wind power and 4 GW of geothermal and solar power during this period.[5] Is this possible? With commitment, sacrifice, and determined leadership, it is.

As usual, California provides a challenging benchmark. Governor Arnold Schwarzenegger embarked the state on an ambitious, many say unrealistic, program to generate 33 percent of the state's electrical power from renewables by the year 2020, up from 12 percent today. Much of the growth will come from new wind farm projects. Again, is this possible? Perhaps so, thanks to a significant convergence of national interests. Environmentalists and scientists are concerned with global warming; however, as the nation struggles to emerge from a serious recession, politicians and economists fear the loss of jobs. In the immediate future the latter problem must take priority. The Obama team hoped to create some 3 to 4 million jobs, although the actual number is in doubt. Trade organizations, such as the American Wind Energy Association (AWEA), believe that their member companies can help meet Obama's goals. Certainly a significant portion of the economic stimulus package, perhaps $25 billion, has assisted wind energy companies, mainly through modification of subsidies and federal loan guarantees. It has resulted in placement of hundreds if not thousands of new wind turbines.[6] Whatever the details, public policy favors renewables, and wind energy would seem on the cusp of significant growth.

There are experts who believe that now is the time for the nation to make a "green America" a national goal. Through his film *An Inconvenient Truth* and his call for a 100 percent renewable-energy nation within ten years, Al Gore has awakened many Americans to the threats climate change poses to the planet. Environmental journalist Thomas Friedman believes the nation hungers for such a national goal. "What kind of

America would you like to see," he asks, "—an America that is spotlighted as the last holdout at international environmental conferences, earning the world's contempt, or a *green America* that is seen as the country most committed—by example—to preserving our environment and the species that inhabit it, earning the world's respect?"[7] Friedman continues: "Yes, the wind has changed direction. . . . So I say we build windmills. I say we lead." Every indication suggests we will. The political climate, the economic situation, and the environmental imperatives indicate that wind energy's star is rising. Many criticize it—and for legitimate reasons—yet the preponderance of arguments favor its growth. A brownout or blackout has a way of redirecting priorities. We may reach a point where the alternative is so unpalatable that wind energy is the most acceptable of unacceptable electrical technologies. If our nonrenewable energy sources diminish and the effects of global warming persist, we will have to call on more and more of the big turbines for help. As a nation, we have been selfish and inflexible, enjoying the fruits of a carbon-based abundance, an abundance that may end sooner than we think. Given that electricity is not optional in our modern world, we must be flexible in how we produce it.

Windfall

CHAPTER 1

How Have We Used Wind and Electricity?

The free benefit of the wind ought not to be denied to any man.
—Herbert of Bury, Suffolk, circa 1180

Some years ago when I first began studying the history of wind energy, I visited the manager of a Palm Springs wind farm (San Gorgonio Pass) to ask a few questions. I told him I was writing the history of wind energy from 1880 to 1980. He replied, with a certain assurance, "Why would you want to do that? Nothing happened before 1980." In a sense he was correct, for the huge turbines and wind farms that dot the landscape today did not exist before 1980. On the other hand, he was like a county commissioner I knew who announced as he grew older that if he couldn't remember it, it did not happen. His perspective of history was rather skewed, and so was that of my wind energy manager. When we view today's wind turbines, we have difficulty comprehending the past. The turbines seem too modern to have a meaningful precedent. Yet, although the technology may seem unrecognizable, the concept of employing the wind to get work done is as old as human civilization.

Origins

To understand something of that past is to have a greater appreciation for the present. Humans have come a long way in our understanding and use of the wind, and this initial chapter provides an idea of the dynamic present by reviewing the distant past. To look first at the energy source: the wind is a function of the relation of the earth and the sun. Wind is a form of solar energy, created when sun-heated air rises and cooler air rushes in to fill the vacuum. In the millions of years of prehistory it accomplished remarkable work. Erosion, which is responsible for much of the land's appearance, is often been attributed to water and ice, but the wind is a significant player too. In the American West, where natural forces are so evident, weathered mushroom-shaped sandstone hoodoo towers, such as at Bryce Canyon, attest to the never-ending work of the wind. Immense concave valleys, as represented by the Laramie Plains in Wyoming, give evidence of its power and longevity. Flying over the West one observes the wrinkled land, the upthrust pinnacles, flat mesas, and broad sagebrush-covered valleys; a forbidding land all shaped by the wind and its cohort, water. One geologist credits the wind with the "exhumation of the Rockies."[1] Now we are determined to capture that same wind, put it to our use, and then release it to continue its movement and its work.

The relationship of wind to human beings is complex, and freighted with mythology. Myths about the origins of things abound among peoples throughout the world. My favorite myth is the Greek story of Odysseus, Homer's hero and wanderer of the Ionian and Aegean seas. On one of his sailing adventures Odysseus received from the god Aeolus, keeper of the winds, a wineskin containing contrary winds. Aeolus warned Odysseus not to untie the wineskin, but of course his curious crew mates did so. An enraged Aeolus stilled the wind, forcing Odysseus and his crew to row their ship on calm seas for six days. "No breeze, no help in sight, by our own folly—six indistinguishable nights and days," lamented Odysseus.[2]

The story of Odysseus and Aeolus reminds us that the historic importance of the wind is linked to exploration. Sail power was the only method to negotiate vast stretches of the globe's oceans and seas. For thousands of years human discovery, conquest, and trade all relied on sailing ships. From ancient times to the middle of the nineteenth century, the bubbling wakes and billowing sails of thousands of ships tell the story of commerce and conquest. With the exception of oared boats—rarely used for long voyages—there was no other power source available. For better or worse, trade, migration, and the exploration and conquest of the world were all dependent on a human partnership with the wind.

When did this alliance begin? Definitive answers are lost in antiquity, but anthropologist Geoffrey Irwin believes that Asians migrated as far as Australia using crude sailing boats some forty thousand years ago. No artifacts have survived, but Irwin remains convinced that Asians explored and settled Australia by means of "voyaging corridors" of no more than one hundred kilometers between land points.[3] Much later, around 1500 B.C.E., Polynesian peoples colonized many Pacific islands, using catamaran sailing craft of forty to fifty feet in length. Without any identifiable technology these deepwater navigators demonstrated a remarkable knowledge of winds and tides. In the Mediterranean region, water transportation was not so sophisticated. The Egyptians, a river people, had developed graceful sailing boats called feluccas by 3100 B.C.E. Some feluccas were quite large and were used for transporting cut granite from upriver to the Cairo area for pyramid building. They floated slowly down on the river current and then used the prevailing northern winds to return to the Luxor quarries.[4] Feluccas still ply the Nile today, often in pursuit of tourist dollars.

This remarkable use of wind power over many centuries has largely disappeared, replaced by steam and diesel engines. Sailing today is a sport or leisure activity, although smaller sailing ships do still ply the seas, and inventors are experimenting with sailing rigs to give an energy assist to large ocean-going vessels. In Germany a company called SkySails is designing and developing a sail that resembles a giant paraglider. Using electronic controls and an automatic retraction system, the sail can range from 100 to 300 meters in the air to catch winds high above the water surface. The company expects a ship equipped with its sail to consume 10 to 35 percent less oil than usual. In January 2008 the first large "sail ship," the MS *Beluga Skysails*, departed northern Germany for Venezuela, arriving in the middle of March. It then continued to the United States and eventually to Norway. When the "kite" was in use, the ship saved 10 to 15 percent in fuel, or $1,000 to $1,500 per day, the pre-industrial energy source of the wind being combined with modern fossil fuel to create an innovative hybrid.[5]

Terrestrial Use

The first terrestrial use of wind is likewise hidden in history. When did the first windmill turn and begin the centuries-long harvest of wind power? One historian who explored the question confessed that such hidden knowledge "is beset with false trails and episodic detours."[6] Supposedly, Hero of Alexandria, a Greek naturalist, constructed a small windmill, providing power to play an organ; if true, this classifies as play rather than work. Not until the tenth century do we have evidence of windmills at work, and that was in the blustery Seistan region of Persia (Iraq). Their turning technology would not be familiar to Europeans. They used sails made of reeds, the sails following a carousel's path around a vertical axis. The Persians employed them to grind corn and wheat or to raise water for irrigation purposes. With time this technology spread east to India, other parts of the Muslim world, and China.[7]

In Europe a different kind of windmill evolved, and it is from the post-mill that our huge turbines of today trace their ancestry. William of Almoner, of Leicester, England, erected the first known post-mill in 1137, almost certainly prompted by studying the mechanics of the popular waterwheel. Historian Terry Reynolds suggests that "very likely the invention of the Western vertical windmill was inspired by the widely used vertical waterwheel."[8] The post-mill quickly spread throughout England and then to Spain, France, Belgium, Holland, Denmark, the German principalities, and the Italian states. As these post-mills multiplied in number and grew in size and technical innovation, they took on a different appearance. Frequently the tower-mill replaced the post-mill. The former featured a four-bladed wheel mounted on a swiveling cap, set atop a stationary tower. The Dutch in particular adopted the tower-mill, which has become synonymous with that nation and is still a national icon. At one point eight hundred to a thousand tower-mills served the power needs of Amsterdam. They pumped water, ground grain, powered sawmills and paper plants, and performed many other rotary tasks. What is most interesting, however, is the cultural acceptance of the windmill. The brightly painted Dutch mills, according to one historian, were often the artistic centerpiece of a village. The mill was commonly a place of public gathering, sometimes for wakes and mourning, sometimes for celebration with the mill draped in "brightly coloured flags and garlands."[9] As we debate the appropriateness of today's huge turbines, we might reflect on the marriage of windmills and culture in Holland. The windmill seemed to express an underlying harmony for the village and a harmony with nature as well. For residents, the mill represented an aesthetic addition to the community; an enhancement to the landscape.

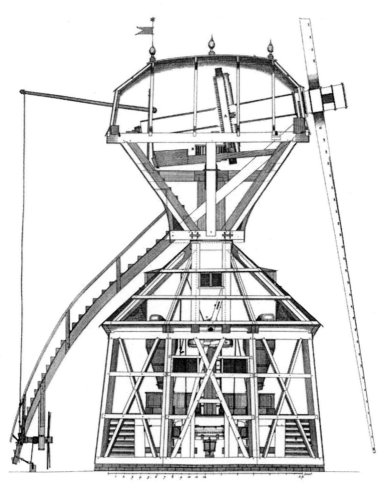

Dutch windmills were large, bold, and beautiful. They combined utility with attractive design and architectural innovation. This scale drawing is from Lacendert van Natrus et al., *Groot Volkomen Moolenboek*, 1734, vol. 1, plate 8. (Courtesy of the Huntington Library, San Marino, California.)

Egalitarian Nature

An aspect of medieval windmills that is often overlooked is their egalitarian nature. Neither solar energy nor wind energy can be controlled by any person or corporation. They are free energy sources, open to all who can figure out how to utilize them. This certainly is one of the principal reasons why windmills thrived in England. Water power was strictly controlled. In England the clergy and the landed gentry owned the streams. Under English common law these riparian owners received all legal rights to the water. Landowners not bordering streams or rivers had no rights. Thus in twelfth-century England, riparian owners controlled the only power source for grain mills, a monopoly no peasant could avoid. These owners had cornered the market, much as the nineteenth-century American railroad companies had control of transportation, charging farmers prohibitive rates to get their crops to market.

One case may illustrate the point. After William of Almoner's post-mill appeared, others followed. Small entrepreneurial farmers owned many of these mills. A farmer named Herbert and his two sons erected a post-mill in the late 1180s. Unfortunately, Herbert had not counted on Abbot Samson, a cleric whose monastery contained a profitable watermill nearby. When he heard of Herbert's windmill, the red-bearded cleric flew into a rage, determined to squelch any competition. He ordered his carpenters to tear down Herbert's windmill. In a desperate plea Herbert won an audience, and during their heated exchange, Herbert proclaimed that "the free benefit of the wind ought not to be denied to any man." His profound statement did no good. Although he promised to grind only his own grain, political power was not on Herbert's side. Abbot Samson did not claim he owned the wind, but he did insist that he had jurisdiction over what could be built within the region. He could, in effect, determine land-use zoning. Samson again ordered that Herbert's mill be destroyed, but before his men could accomplish the task, Herbert's sons dismantled the post-mill.[10] Since this farmer's time the idea of the wind as a free energy source has been a compelling attraction. "The free benefit of the wind" has been a lure for creative entrepreneurs. The story of Herbert provides a wonderful example of how nature does not play favorites, dispensing this energy source without consideration of human class or needs.

Throughout American history independent farmers and ranchers have erected windmills for domestic water use and also to make arid ranch land usable. By tapping ground water, these agrarians became independent of surface streams, which were often controlled by large ranching interests. Settlers' satisfaction in using a free, abundant source of energy had little to do with environmental concerns and more to do with Benjamin Franklin's dictum: "Waste not, want not." Wind energy has always been associated with individual freedom. Many people savor independence from central-ized power, whether it relates to government, religion and spiritual life, or energy. In our contemporary world, energy self-reliance often means freedom from a utility company. Seeing a windmill turning provided an undeniable satisfaction to the owner, for the fuel was free. No wonder the American windmill became an icon of independence and freedom, and small wonder that today thousands of landowners continue the tradition of independent, decentralized electrical production.

The focus of this study is the big commercial wind turbines, but we ought not to forget the thousands upon thousands of small turbine own-ers, who, like Herbert, simply want a modest, simple, reliable power-producing turbine that gives them the satisfaction of independence. For Herbert and his family, being free of Samson would surely make life more enjoyable. Essentially, they wanted no part of the restriction and injustice that came from a powerful institution. Today, there are many Americans like Herbert. In chapter 8, "Small Turbines and Appropriate Technology," I examine this aspect of wind energy.

Windmills Spread

Small post-mills, then, quickly crossed the English Channel to thrive in France, Spain, the German states, and Denmark. Their spread was syn-onymous with rising protests against lordly monopoly. As a European his-torian put it, the windmills were "established in the conditions of freedom that opened with the growth of cities, and established a further breach in the lords' energy monopolies."[11] The work that windmills accomplished had profound social consequences. The eminent medieval historian Lynn White believed that "the chief glory of the later Middle Ages was not its cathedrals or its epics of its scholasticism; it was the building for the first time in history of a complex civilization which rested not on the backs of sweating slaves or coolies but primarily on non-human power."[12] White underscores an idea we often forget: options expanded in the world of work and energy, and windmills were part of that revolution. By the late nineteenth century workers had erected approximately one hundred thou-sand windmills across Europe. And according to Lewis Mumford, "the greatest technical progress came about in regions that had abundant sup-plies of wind and water."[13] Thus the basic technology of the windmill as a tool of energy may be found in pre-industrial Europe. Furthermore, the consequence of spinning windmills was immense. They harnessed power and accomplished work, and in the process they freed human beings from drudgery.

Coming to America

As thousands of colonists came to settle the Atlantic seaboard of North America, the windmill accompanied them, but only in a limited way. The European windmill was too large, cumbersome, and expensive to survive, let alone thrive. It required constant human attention, and in this new environment labor was scarce, even if land was abundant. Although colonists erected a few windmills in Jamestown, New Amsterdam (New York), and the Cape Cod area, they depended primarily on water power. The colonists used wind power at sea, of course, but on land it was human and animal labor, wood, and water that got things done.

Clearly the physical environment determined how early American settlers from Europe might meet their energy needs. It was in the American West that wind energy eventually found a significant role. West of the hundredth meridian the wind was strong, and frequently destructive. It was often a topic of conversation, rarely in a positive way. Ever-present winds knocked down trees, flattened gardens, blew off roofs, and carried off things that were not battened down. Tornados are still a source of terror in the Midwest. Less fearsome but more destructive were the 1930s Dust Bowl winds, which stripped the land of its soil and the people of their livelihood. Wind stories abound in the West. In Laramie, Wyoming, where I lived for fifteen years, that standing joke was that "one day the wind stopped blowing and everyone fell down." Just about every western town has a variation on that theme.

And yet the constant westerly winds did perform useful work. American windmills had several purposes. They provided water for homesteads, for livestock, and for the thirsty steam locomotives of the Union Pacific. They changed the western rangelands, creating stock wells and small vernal oases. Innovative Americans produced a great variety of styles and brands of windmills, which dotted the Great Plains. They all had certain characteristics in common. They were small, light, movable, self-regulating, and inexpensive. There is no way to estimate their numbers accurately, but some authorities have offered a figure of more than 6 million.[14] Today some one hundred thousand of these mills continue their work; if you look closely, a little windmill is often still to be seen doing its job among the huge new wind turbines.[15] Ranch windmills have become objects of nostalgia. They evoke an earlier, bucolic time, in part because they employ technology at what a majority of Americans perceive as an acceptable level of disruption to landscape and nature. They seem to complement rather than corrupt the scenery.

The Marriage of Wind and Electricity

Unquestionably the windmill would be significant in the nation's history even if electricity had not transformed how Americans live. Yet it is the marriage of wind power and electricity that promises to impact our energy future. This is the best hope for a significant portion of the nation's energy needs to come from a renewable resource.

Scientific understanding of electricity dates to the eighteenth century, when Benjamin Franklin and others conducted experiments and published their findings. However, the scientists could not bridge the theoretical and the practical for this new power until 1844, when Samuel Morse successfully transmitted a telegram message across a distance of thirty miles. In the years following the Civil War the uses of electricity multiplied. Historian Paul Valéry suggested that the decades from 1870 to 1900 represented "the conquest of the earth" by electricity.[16] Perhaps this was an exaggeration, but not by much. The creation, distribution, and consumption of electricity in the final decades of the nineteenth century captivated the scientific community. And although we might surmise that wind-generated electricity played no role, we would be wrong.

Some of the great industrialists and pioneers of electricity found the power of the wind a fascinating possibility. Andrew Carnegie, lamenting the depletion of coal and iron ore, reminded the 1908 Conference of Governors that other forms of energy should be developed. "The sun-motor still runs . . . and may yet be made to produce power through solar engines," he suggested.[17] He did not define "solar engines," but surely the windmill would be one.

Two giants of the scientific world, Thomas Edison and Nikola Tesla, fought fiercely over the practicality of transmitting electrons as direct current (DC) or alternating current (AC). Tesla won the argument, and after his partnering with George Westinghouse, alternating current became standard. Edison and Tesla did agree, however, on the potential of wind power. In 1901 Edison embraced the idea of wind electricity for rural inhabitants, and by 1911 he had made drawings of a windmill to power a cluster of four to six homes. He approached a manufacturer to produce a prototype, but nothing came of it. Tesla, like many people, was already concerned with depletion of fossil fuels. He favored placing wind turbines atop homes and buildings to pump water, power elevators, and cool homes in the summer and heat them in the winter.[18] Even earlier, in 1881, the great English scientist Lord Kelvin subscribed to wind energy research and development. He was worried about England's dependence on coal. Why not employ a free, abundant energy source if it was available? He did not believe it "utterly chimerical to think of wind superceding coal in some places for a very important part of its present duty—that of giving light."[19]

Whereas numerous scientists and inventors speculated about the benefits of wind energy, one man did something about it. In 1886 Charles Brush, a Cleveland scientist with a practical, inventive bent, decided to build a large wind turbine on his spacious five-acre backyard. His was the first large wind turbine in the world, and it was the culmination of an idea that had fascinated him from boyhood. Unlike Edison or Tesla, he did not anticipate that his turbine would change the way he and his neighbors (John D. Rockefeller, among others) received electricity. He simply wanted to experiment with wind-produced electricity to energize his own laboratory motors and light his mansion. He had already made a fortune through his arc light system, purchased by many cities throughout the nation to illuminate their downtown streets.

By 1888 people strolling along Cleveland's fashionable Euclid Avenue might have been dumfounded to see a 60-foot tower with a rotor, or wheel, of 144 blades reaching another 20 feet into the air. The rig weighed 80,000 pounds. A complicated system of gears and pulleys assured that the dynamo, or generator, made fifty revolutions to one of the rotor. The generator fed an elaborate system of batteries in the basement of Brush's mansion. It was an amazing machine that caused the editor of *Scientific American* to rhapsodize: "As an example of thoroughgoing work it cannot be excelled."[20]

But it did not last long. At the turn of the century Brush and his wealthy Cleveland neighbors hooked into central power, rendering his wind dynamo obsolete. Yet Brush maintained the novel machine, and as he later recalled, it "was built to go for twenty years and it never failed to keep the batteries charged until I took the sails down in 1908."[21] From the perspective of reliability we must note that perhaps the major accomplishment of Brush was not so much the building of the dynamo but the fact that it

Charles Brush built this huge wind turbine in 1886 in his Cleveland back-yard. It was the first large, functional wind-to-electric turbine in the world. To gain a sense of the machine's size, note the garden worker cutting the lawn to the right of the windmill. It ran successfully for more than twenty years, but unfortunately was eventually dismantled and destroyed. (Courtesy of Western Reserve Historical Society, Cleveland, Ohio.)

operated for twenty years without any significant down time. That was a meaningful accomplishment, and a record modern wind farm managers would envy.

At Brush's death in 1929 all that remained of the wind dynamo was the 60-foot tower, standing forlornly on the spacious property. Its final fate was a sad one. A representative of Henry Ford's antiquarian interests wanted to move the Brush windmill to the Museum of American Antiquities and Edison Institute for Technology, both in Dearborn, Michigan, where Ford's staff would restore the turbine and place it on permanent display. But in response a Cleveland committee protested, and the city spurned Ford's offer. The committee arranged to place the windmill on the Case Western Reserve University campus. However, as the Depression of 1929 deepened, money grew scarce, commitment weakened, and the committee finally authorized storage of the wind dynamo in an old warehouse. In the late 1950s a zealous executive for the Clevite Company, looking for storage space, ordered the windmill tower removed and destroyed.[22] Thus the world's first large wind turbine did not survive except in a photograph. The loss may be attributed to an efficient business manager, but it is also reflective of the nation's indifference to the possibilities of wind energy.

A Practical Use

Although Brush was always careful to register his inventions, he never pursued a patent on his dynamo. He knew it was one-of-a-kind and far too expensive and complicated for commercial use. Thus while Europeans, particularly the Danish professor Poul la Cour, continued to experiment and gain knowledge about wind electricity, nothing happened in the United States. Many rural Americans were suspicious of new technology, particularly electricity. They shared the concern the *Rural New-Yorker* editor expressed when he warned readers that electricity "is a dangerous servant liable to strike its master dead at a single unguarded touch."[23] Wind electricity represented a technological enigma. It was deceptively simple, yet it was mysterious and provoked fear. The author of the first book devoted to wind turbines regretted that "there is every prejudice against wind-power on account of its uncertainty, against which even its inexpensiveness has not been able to contend."[24]

As a result of such trepidation and skepticism the decades between 1890 and 1920 represented the pause between invention and application. Furthermore, American inventors and engineers focused their efforts on interdependent urban electrical systems, not *individual* energy needs. One vast area of the country, however, was well suited for single wind energy units. Rural America, seemingly detached from modernization, had no electrical system and was unlikely to get one in the foreseeable future. Compared to their urban neighbors, farmers lived in a state of technological deprivation.

A Reliable Machine

However, help was on the way. During and after World War I, engineers and scientists made major advances in their understanding of aerodynamics. Two brothers, living on their parents' ranch in eastern Montana, profited from this new knowledge of airplane propellers and electricity. Joe and Marcellus Jacobs were tinkerers with mechanical aptitude. They could wrestle with problems and think out solutions for themselves. The boys generated electricity from a small gasoline Delco generator. But fuel was dear. Hauling gasoline by wagon from the tiny town of Wolf Point to power a generator took three days and required crossing the Missouri River. There had to be a better way to enjoy the benefits of electricity. Why not use the free, abundant fuel of the wind? After many experiments the boys succeeded in building a three-bladed prototype turbine that worked beautifully. They sold a few to their neighbors, then created the Jacobs Wind Electric Company and soon moved the whole operation to Minneapolis, Minnesota.

By 1930 the Jacobs boys had perfected their turbine, a three-bladed rotor that, aside from size, does not look much different from the standard Danish turbine of today. Other wind-electric turbines were available, but the Jacobs model was superior in both power and reliability. Clearly they were ahead of their time, proving that one did not need an engineering degree to build a quality turbine. Between 1930 and 1955 the company built approximately thirty thousand units. They were legendary for their reliability. When Richard Byrd returned to his "Little America" outpost in Antarctica in 1933, he took along a Jacobs turbine and installed it on a 70-foot tower. When Byrd's son returned to the deserted site in 1947, he found the turbine blades still spinning. In 1955 one of the 1933 veterans removed the blades, which would otherwise soon be buried under ice. It was the ability of the Jacobs turbine to run day in and day out—the reliability factor—that made it such an exceptional machine. It was a model of efficiency, and a few are still in use today.[25]

Killing the Wind Industry

Although the Jacobs turbine was a fine machine, it could not overcome
the indifference of the federal government regarding wind energy. The
Rural Electrification Administration (REA), through its many local cooperatives, strung electric lines (called the "hi-line") to rural America.
Admittedly, this New Deal agency transformed country life, bringing electricity to thousands who were without it. However, it also killed the wind-
electric companies. By 1956 no company had survived. We know today
that the fast-growing wind industry would face difficulty without federal
subsidies, but in the 1930s, the industry was on its own. It received neither
federal assistance nor research support until 1973. We had an opportunity in the 1940s to have a centralized system with room for independent
"stand-alone" systems, but the opportunity was squandered by rather myopic bureaucrats.

Lost to history are the stories of those rural people who already pos-
sessed wind-generated electricity and had no desire to hook up to the REA
lines. REA power was a blessing in freeing rural people from drudgery,
as had windmills in medieval Europe. Yet viewed from another perspective, central power merely added another nail in the coffin of debt and
dependency that plagued the American farmer throughout the twentieth
century. As historian David E. Nye has suggested, "neither big government nor advanced technology was necessarily consistent with maximiz-
ing democracy or preserving the small farmer, and rural electrification
certainly would not recognize the agrarian dream of a decentralized blend
of technology and Jeffersonianism."[26] Nye believed that technology was a
double-edged sword for farmers. It freed them from many onerous tasks,
but the technology cost money, and for many farmers it led to increased
debt and eventual foreclosure.

When R. E. Weinig testified before Congress in 1945, it was with the
hope of maintaining a modified decentralized blend. According to this
vice president and general manager of the successful Wincharger Corporation, the REA should consider support of wind turbines in isolated areas.
In the sparsely settled West, statistics showed that there were half a million
farms or ranches that averaged only 0.6 farms to the square mile. This
average was less than one-fourth the necessary density to make an REA hi-
line feasible. The agency's engineers maintained that 2.5 to 3.0 hookups
per mile were necessary to make an REA cooperative venture successful.

Therefore, Weinig reasoned, since stringing wires to these remote ranches made no economic sense, why not encourage decentralization through individual electric generating plants? He suggested that the REA bill in question, H.R. 1742, be amended to provide ten-year, low-interest loans to qualifying farmers and ranchers for individual energy plants.[27]

Needless to say, Weinig's proposal fell on deaf ears. Bureaucrats of the REA, the Tennessee Valley Authority, and the utility companies were wedded to central power systems. They had no patience for stand-alone systems, which seemed to them a primitive idea. As the REA continued to develop rural lines, often ignoring its own hookup density standards, the wind energy companies closed their doors.

Creative Experimentation

While small turbine producers struggled, one brilliant, ambitious engineer constructed a giant wind turbine atop a hill in the Green Mountains of Vermont. His name was Palmer Putnam, the well-heeled child of a prominent publishing family. When Putnam built a cottage on Cape Cod in 1934 he "found both the winds and the electric rates surprisingly high." Wind energy seemed natural, but it was impossible to hook into the local utility since no device existed to transform the turbines' DC power to the AC system of the utility. He soon invented one. Then the economy of scale captured his imagination. By 1940 he had assembled a remarkable team of academic and private company experts, all under the financial umbrella of S. Morgan Smith Company, a firm that specialized in water turbines but was hoping to branch out. Under Putnam's leadership and without federal subsidies, the team built and then assembled a spectacular 1.5-megawatt (MW) turbine on Grandpa's Knob in the Green Mountains of Vermont.

The building and operation of this seminal wind turbine is a story in itself, and Putnam wrote a book about it.[28] The turbine operated successfully for sixteen months, beginning production on October 19, 1941. When it broke down in February 1943, the war effort made both parts and labor near impossible to obtain. Still, Putnam got it up and running again until March 26, 1945, when it suffered a catastrophic accident. As a result of metal fatigue, one of the huge blades broke free of the rotor and careened down the mountainside. Putnam promised he would address the issue of metal fatigue and soon have the turbine rebuilt, but that never happened.[29] The S. Morgan Smith Company could not fund further such a questionable experiment. Moreover, engineers and planners saw little

In 1943 the brilliant engineer Palmer Putnam designed this massive wind turbine and erected it on Grandpa's Knob in the Green Mountains of Vermont. Workers built the turbine from the designs of a number of co-operating engineers and scientists. It was largely funded by the S. Morgan Smith Company, manufacturers of hydraulic turbines, with no federal subsidy. After sixteen months of successful operation, the 1.5 megawatt turbine lost a blade. Wartime shortages prevented Putnam from repairing his machine. Besides its impressive size, the turbine was distinguished in having the capacity to transform DC power to AC, making the electricity compatible with the grid of the Central Vermont Public Service Corporation. (Photo by Don Guy, courtesy of Don Guy, Jr., Sarasota, Florida.)

future in wind power. Atomic energy had appeared on the horizon, and with it came breathless claims about producing electricity so cheaply that it would not require a meter.

In the wake of the Putnam turbine, a few engineers kept interest alive, but just barely. From 1945 to 1973 wind energy was an interesting idea, but neither private industry nor the government had faith in its practicality. Yet the research continued. Percy Thomas, a high-ranking engineer with the Federal Power Commission, had followed the Smith-Putnam turbine success with fascination. He was soon obsessed by the engineering profession's fondness for the economy of scale. Accordingly, he designed an immense wind plant featuring two nacelles and rotors. It would be rated at 7.5 MW and would rest on six V-shaped legs resembling the Eiffel Tower, with blades reaching higher than the Washington Monument. Thomas envisioned three of these 7.5 MW plants, which would work in conjunction with a hydroelectric system. Alongside these mighty wind turbines, water could be stored in reservoirs for use when the wind ceased to blow. By 1950 Thomas went before Congress to ask for money to build a demonstration turbine. In an era of McCarthyism and the Cold War, Thomas and Department of the Interior Assistant Director William Warne argued the advantages of decentralized electrical power plants that would be impossible to sabotage. But Congress remained unconvinced; Thomas's engineering designs went untried.[30]

A Nuclear Engineer Converts to Wind

The 1960s also witnessed no wind energy progress. American scientists and engineers focused on nuclear energy and fossil fuels to meet the nation's electricity needs. In the lead was Admiral Hyman Rickover, a poster child for nuclear energy. Working under him was Captain William Heronemus, an Annapolis graduate who helped design the first nuclear submarine, the USS *Nautilus*. Yet in time, Heronemus recognized the costs and dangers associated with nuclear energy. His cohorts were sweeping the true price of nuclear power under the rug and minimizing the problem of radioactive waste. He became a critic, considering nuclear energy a costly, dangerous, and foolish way to power our homes. In a jocular moment Heronemus suggested that utility executives should be required to spend their retirement years inside a shut-down nuclear reactor. In what was almost an epiphany, he renounced nuclear energy and embraced renewable sources, specifically the wind.

He retired from the navy in 1965 to continue his research in an academic setting. As professor emeritus at the University of Massachusetts, he established the renewable energy program, training and inspiring many young engineers to think about wind energy when it was totally overlooked. His research showed that large-scale wind development was feasible in the United States. As a member of the National Science Foundation (NSF) and National Aeronautics and Space Administration (NASA) evaluation board in the early 1970s he questioned funding nuclear fission research. He was the first to argue that a fair share of research funds must go to renewables such as wind energy. For his era, this accomplished scientist adopted a unique and courageous position.[31]

When William Heronemus died in 2002, his friend and colleague at the University of Massachusetts, Jon McGowan, declared his most important legacy to be "the contributions of the numerous students who worked under his supervision."[32] Perhaps the most successful was Forest "Woody" Stoddard. With his mentor's blessing, Stoddard applied for a grant to develop a 25-kilowatt (kW) turbine that could be connected to the grid. Lou Divone, an MIT graduate who shared Heronemus's passion for wind energy and worked at the NSF, fought to see it funded. On the surface, this project seemed like reinventing the wheel. However the only turbine left standing after the 1950s "massacre" of windmill companies was an inappropriate Australian model. Finally, after lavish funding of research and development in the pork-barrel world of nuclear science (in the range of $50 billion), the NSF threw wind energy advocates a tiny bone. It was not much, considering that scientific knowledge of wind energy was practically nonexistent.

An Energy Crisis Exposes a Flawed Policy

Attitudes changed in 1973, when for the first time in American history the nation faced a painful energy crisis. The oil embargo initiated by the Organization of Petroleum Exporting Countries (OPEC) resulted in long lines of cars at gasoline stations and escalating prices for all fossil fuels. It underscored the weakness and dependence of a nation that presumed extravagant energy use was a national right. Overnight, Congress and the country began looking to renewable energy for part of the solution.

Some of the resulting federal programs are described in chapter 2; suffice it here to mention one project with which I was familiar. On Highway 30 north of Laramie, Wyoming, sits the tiny windswept railroad town of

Medicine Bow, known only for its role in literature as the cow town in Owen Wister's path-breaking novel *The Virginian*. In 1979 the Bureau of Reclamation and the Department of Energy arrived with federal dollars to spend. They proposed creating a wind farm of fifty turbines (representing 100 MW), producing "enough power to serve 67,000 homes."[33] Engineers planned to integrate the turbines into the great power-generating dams of the Colorado River, primarily Flaming Gorge and Glen Canyon. It was an admirable idea, but wind technology was not up to the task. To begin the Medicine Bow project the agencies funded two turbines: a Boeing MOD 2 (2.5 MW) and a Hamilton Standard WTS-4 (4 MW), the latter touted as the world's largest windmill. The two turbines were up and running by 1982, impressive, eye-catching machines on the vast open landscape. The blades of the MOD 2 reached 350 feet, while the WTS-4 towered to 391. For four years the two turbines produced electricity, but only occasionally. Plagued by problems from the start, the Boeing MOD 2 ended its brief life with a burned-out main bearing. When bureau officials received a re-pair estimate of $1.5 million, they tried to sell the turbine. There were no takers, and finally a scrap dealer bought the unit for $13,000, dynamited the tower, and hauled it away. This was not an unusual performance for Boeing units; we can be grateful the company builds better airplanes than turbines.

The WTS-4 operated for four years but came to a halt when a loosened bolt ground up inside the generator. Although the local population thought it might serve as a tourist attraction in a land of few features, it was des-tined for removal. Bill Young, an electrical engineer from Casper, bought the $10 million, 4 MW turbine from the bureau for $20,000, repaired it, and occasionally ran it. In 1994, in a heavy wind, the pitch control failed, causing one of the blades to hit the tower, sheer off, and crash to earth. The next day in 60-mile-an-hour winds the four-ton platform around the nacelle fell to the ground, and by the end of the day the base of the tower was littered with debris. Needless to say, the WTS-4 was finished.[34] Similar stories plagued MOD experiments elsewhere.

California Green

In spite of the large wind turbines' miserable performance, the federal government continued its commitment to renewable resources. Interest in and expansion of wind turbines arrived not from advanced technology but rather through federal law. In 1978 Congress enacted the Public Utility

Large 2.5 MW turbines were plagued with mechanical problems in the 1980s. When this Boeing MOD 2 at Medicine Bow, Wyoming, required repairs amounting to $1.5 million, the Bureau of Reclamation sold it for $13,000 to a scrap dealer, who dynamited the tower and hauled it away. Boeing Corporation built better airplanes than wind turbines. (Courtesy of the *Casper Star-Tribune*, Casper, Wyoming.)

Regulatory and Policies Act (PURPA), a law that obliterated the utility companies' monopoly of energy production, throwing open the doors of free enterprise. A massively complex act, it contained a crucial section mandating that utility companies must accept independently produced electricity, such as wind power, on their grid and must pay what are known as "avoided costs." How this vital mandate effectively shattered the monopoly of utility companies is examined more fully in chapter 5.

In California, PURPA combined with federal and state government incentives to create an attractive business opportunity. It provided the catalyst to move California in the direction that Governor Jerry Brown (1975–83) was determined to go. Always on the cutting edge of environmental progress, Governor Brown had set up the Office of Appropriate Technology and had hired environmentalists Sim Van der Ryn and Ty Cashman to urge the state legislature to grant tax credits, augmenting incentives offered by the federal government. PURPA, the tax credits, plus other spurs created the financial climate that kindled the remarkable growth of the wind energy business in the early 1980s.[35] The California legislature then delivered the Mello Act, which set forth an objective that 1 percent of the state's energy would be generated by wind power by the year 2000. Although the goal was not mandated, it can be considered the first state renewable energy portfolio target. California also conducted extensive wind surveys with encouraging results. Wind energy had a governmental advocate!

All these factors were present when California experienced a "wind boom" that has often been compared to the Gold Rush of 1848–49. Soon wind turbines were at work at Altamont, Tehachapi, and San Gorgonio passes. By 1985 workers had installed 12,553 wind generators with a capacity of 911 MW. The generators had a global persona, with representative turbines from Germany, Belgium, Denmark, the Netherlands, Ireland, Scotland, and Japan covering the hillsides. They represented 96 percent of existing wind capacity in the world. For a time the construction sites were reminiscent of the old gold-mining camps or an oil boom town. In the town of Tehachapi, more German, Danish, and Italian were heard than English. Cowboys, Germans, and Danes argued and fought in the bars. Public hearings turned into shouting matches. Fraud and deception became a way of life. Paul Gipe remembers that as an exhilarating yet maddening five years: "You had to be there. It was like the Wild West." But this was not necessarily bad. He compared it to a community effort by the whole town. They were "building something together—like the Amish at a barn raising. It was a heady experience."[36]

The results of this great boom were mixed. Manufacturers and developers tested their turbines in a realistic situation. They learned much. Installation developers discovered that the big turbines produced by such companies as Boeing, Alcoa, Bendix, and General Electric were so unreliable and fly-by-night that in spite of incentives, these participants left the field. Clearly the successful turbines were the Danish models, heavy yet reliable. Within a short time, some 55 percent of installed turbines were from the diminutive Scandinavian country, rural in character but smart with wind energy.

The state of California realized not only how to encourage this new industry but how to regulate it. The state and the developers learned that in spite of the environmental benefits, not every community welcomed the turbines in the proximity of their towns or viewscapes. Particularly in the Palm Springs area (San Gorgonio Pass), protests and the not-in-my-backyard (NIMBY) response resulted in heated exchanges and, eventually, regulation.

California Bust

When a boom comes, a bust is likely to follow. The public soon found out that wind turbines did not represent the economic or environmental panacea they had hoped to see. Few if any of the wind companies paid investors off in cash. Yet the tax incentives were so attractive that advertisements for one company, Oak Creek Energy Systems, told investors "it may not matter which wind turbine developer you choose [since] the tax savings nearly total the entire cost of your wind turbine."[37]

In truth, many of the turbines did not work at all, and yet the investor would not lose money. A wealthy friend of mine invested $50,000, but his turbine (an HMZ Windmaster manufactured in Belgium) was a miserable failure, and he never received a dime in dividends. Yet he lost little money. The tax write-offs were so generous that they could guarantee an investor against loss. The wind energy companies profited as well from generous subsidies. Perhaps the most ludicrous was that federal regulators paid subsidies on the rated capacity of a turbine, not the *energy production* of that turbine. In effect, a turbine did not actually have to produce one kilowatt of power to receive generous subsidies.

When federal assistance ended on January 1, 1986, the industry fell into the doldrums. American companies such as Zond, Fayette, Carter, and FloWind found themselves on the brink of bankruptcy. Close on the heels of bankruptcy and contract cancellations came court cases. Cities sued federal agencies, investors sued wind farm companies, wind farm companies sued manufacturers, and manufacturers sued insurance companies. It was a hard time, and the American public became convinced that wind energy was nothing but a scam and a tax write-off. Freeway travelers along Interstate 10 viewed the Cabezon wind park, for instance, where only about 25 percent of the turbines worked. Broken turbines formed an unhappy eyesore of twisted blades and fallen debris: one reporter saw it as more like a war zone than a wind park.[38] Thousands of wind turbines marred the landscape. Most people were willing to live with the turbines' landscape impact, but only if they worked.

The wind energy industry lost public confidence and a lot of goodwill through participation in unworthy experiments and erection of turbines for avarice rather than performance. Inefficiency, corruption, and unwarranted profit seem to go hand in hand with rapid change and an infrastructure based on need and greed. After decades of dependence on oil and neglect of alternative methods of energy production, the government had suddenly reversed its policy. One wonders if the wind energy could have emerged in any other way than it did: not as a beautifully formed child but rather as a flailing teenager.

Coming of Age

With the new century has come transformation into an attractive mature phase. Problems notwithstanding, it is now evident that significant electricity can be created in a benign, nonpolluting, renewable way that does not contribute to global warming. Problems of landscape aesthetics, avian mortality, and noise exist and will continue to be addressed in coming decades. One wind turbine design became standard: the Danish three-bladed unit. We can hope for alternative options in the future. In regard to size, in the 1990s most experts agreed that the medium-sized turbines of 300 kW to 1 MW would be most practical, but as I explore in chapter 2, that has changed.

Most important, other states have picked up the California experiment. Montana, Minnesota, Texas, and Iowa have all developed their own programs with federal help. Today almost every state except those with inadequate wind (the southern states) has a wind energy program to encourage private or public development, and many have renewable energy portfolios, mandating that from 5 percent to 30 percent of electrical energy must eventually be produced from renewable resources. The wind currently provides the source of most renewable power.

In the pre-industrial world, wind energy was a leading producer of power to perform work. Although we know that wind turbines today will not be dominant and may never produce more than 25 percent of our electrical power needs, one cannot help but marvel at how significant a place this pre-industrial power source now occupies in our post-industrial world.

CHAPTER 2

How Have These Large Turbines Evolved?

Everyone is moving to bigger turbines.
—Gamesa representative, 2008

Texas is wide open for business.
—Governor Rick Perry, 2008

Wind turbines thrive on superlatives. None these days are merely large; they are massive.

If the Cape Wind project on Nantucket Sound is approved by a complicated set of government agencies, Energy Management Incorporated will install 130 General Electric 3.6 MW turbines in the shallow waters of Horseshoe Shoal. Although some six miles from land, they will be visible, a result of their immensity. The new behemoths represent the largest turbines commercially produced in the United States. The blades will reach 440 feet in height.[1] If they perform as expected, the new turbines will create a significant amount of energy.

Just how much energy? The answer is not easy. One difficulty is that the wind is not constant, and although engineers tell us the optimum capacity of the turbines, we can make only a calculated guess at efficiency. Another problem is that wind farm developers like to estimate the power production in terms of houses served. Usually they exaggerate. The accompanying chart shows how to figure the approximate number of homes that will be supplied with wind power when the Cape Wind project moves forward:

Average house kilowatt-hours use per month	2,000
Months per year	12
Average house kilowatt-hours use per year	24,000
Megawatts per turbine per hour	3.6
Megawatts to kilowatts per hour	3,600
Number of hours per year	8,760
Kilowatt-hours per turbine per year at 100%	31,536,000
Efficiency of each turbine figured at 25% (0.25)	7,884,000

7,884,000 kilowatt-hours divided by 24,000 = 3,285, the approximate number of houses that one 3.6 MW turbine will supply with power.

3,285 multiplied by 130 (the projected number of turbines) = 427,050 homes served.[2]

Such figures are impressive. It was not so long ago (1946) that a home-sized wind turbine could be purchased in the Sears Roebuck catalog for $98.80. It would power a Silvertone radio and a few lights.[3] Obviously, the definition of a wind turbine has changed. A new turbine, which can supply electricity for hundreds of modern homes, gives the lie to those who argue that wind power is a hoax only fabricating tax breaks for the utilities and wealthy inventors. The industry would be better served, however, if members were more candid in their estimates. For many years wind energy companies, such as Kenetech, touted that their turbines would produce electricity at 5 cents per kilowatt-hour (kWh), but they never came close. Today the efficiency factor is still a subjective variable in spite of measuring advances. I estimated that the Cape Wind turbines will generate electricity at 25 percent of their rated capacity. That seems to me reasonable, but the American Wind Energy Association (AWEA) prefers a 32 percent figure, which is probably overly optimistic.

Fanciful Ideas

But the questions here are how contemporary turbines grew so large and whether they will grow larger. Wind pioneers have always recognized the idea of economy of scale. The *Scientific American* of the 1880s, a weekly magazine that followed the progress of emerging technology in the United States, reported on some ideas that challenge our perceptions of economy of scale or perhaps represent the inventor's illusions of grandeur. One rather bizarre plan involved 1,600 windmills installed to lift a 24-million-pound weight (1,200 tons), which would then be slowly lowered to provide power for a small factory. Another fanciful American inventor suggested that wind turbines could be employed to force down slowly a huge spring, which would "then drive machinery by [its] recoil." The editors of the *Scientific American* noted that this would create a "fearfully dangerous instrument."[4] Less alarming and more practical, one inventor designed a multiple windmill array to raise water to a reservoir and then release it as needed for power production, a storage and generating method used today. In the late nineteenth century we were not technologically prepared for these futuristic ideas using many wind turbines, yet we can see that inventive Americans were imagining on a scale far beyond the individual farm.

A modern turbine is massive, and a good deal of the structure is below ground. Here workers have constructed an intricate rebar web in preparation for pouring concrete. Critics worry about restoration of the earth when the turbine is decommissioned in twenty to thirty years.

Charles Brush was the first American to take his ideas and plans and actually *build* a turbine of a size that anticipated the turbine of today. His "wind dynamo," as described in chapter 1, was remarkable and certainly pushed the envelope in size, although not in design. Its dimensions were impressive in an era when Americans were familiar only with small water pumpers. Like many originators, he was well ahead of his time. He understood that he was building a dynamo but also a dinosaur. There was no market for such large wind turbines. Central power was coming to Cleveland and cities across the nation. Inventors such as Thomas Edison and George Westinghouse, and Brush as well, were busy creating central power systems that co-opted any interest in wind turbines. By 1900 a complex labyrinth of generators, wires, and poles defined the cityscape of any progressive municipality. The wind dynamo would not be replicated, for there was only one person in the country—Brush—who had both the wealth and knowledge to keep it in operation. A *Scientific American* reporter was almost prescient when he noted that although the wind was free, Brush's plant was not. "On the contrary," he wrote, "the cost of the plant is so great as to more than offset the cheapness of the motive power."[5] Brush agreed with the reporter's assessment. He understood the limitations of his invention, seeing it as an apparatus that was useful only for his inventions and experiments. He certainly never realized the potential of wind energy, nor did he envision that his wind dynamo would create for him an honored position in the history of technology.[6]

The Turbine on Grandpa's Knob

Although there were thousands of farm turbines in use by the 1930s, they were truly midgets in size. As described in chapter 1, the second large turbine was at Grandpa's Knob and came from the imagination of Palmer Putnam, an engineer endowed with an imaginative flair. We should underscore two features of this monumental turbine. First, its size eclipsed anything built to that date and for thirty years into the future. The capacity of 1 MW at a wind speed of 30 mph and 1.5 MW in higher winds far surpassed anything built before the 1970s. It anticipated the power of today's turbines and stretched the imagination of engineers of the era. Second, the turbine's converter synchronized its DC power to accommodate the AC grid. This was a momentous development in the manipulation of electricity. Without it the modern wind plant could not exist. It was perhaps Putnam's greatest accomplishment.

However, Putnam did not build a graceful wind turbine. It was huge in size but ugly in design. The talented team that built the turbine focused on creative engineering, not elegant compatibility. The tower successfully supported the turbine, but its lattice design evoked an oil derrick, functional but hardly attractive. The two blades were rectangular and squarish, with none of the feathering and tapering of a modern blade. An actual wind farm of ten or twenty such turbines would be an unimaginable blight on the landscape. In fairness to Putnam and his engineering team, aesthetics were far down the list of their concerns.

Giant Wind Plants of Today

Neither the world nor the United States in 1945 was prepared to duplicate or advance the Grandpa's Knob experiment. Engineers might study Putnam's book and dream of huge turbines, as Percy Thomas did, but the idea seemed far-fetched. Wind energy escaped the notice of venture capitalists and utility companies in the 1950s and 1960s. The federal government would have to be involved, but it was not interested either.

In the early 1970s, long gas lines, politics, conservation, and a new environmental awareness converged to shift American thinking. For the first time, the nation questioned extravagant use of energy as a national right. For the first time, the federal government could not ignore the critical energy crisis. President Richard Nixon, soon to face his own crisis, fell back on coal and nuclear power for solutions. Yet he recognized that renewable energy did exist and that "solar energy holds great promise as a potentially limitless source of clean energy"; he tripled his solar energy budget to $12 million.[7] This modest appropriation swelled with passage of the Solar Energy Research Act of 1974. The act established the Solar Energy Research Institute (SERI), which was empowered to call on the technical expertise of the NSF, NASA, the Department of Agriculture, and other agencies. All of a sudden research facilities such as Cleveland's Lewis Research Center, administered by NASA, "discovered" wind energy.

The object of their combined effort centered on building a large, reliable wind turbine capable of commercial use. To do that the Department of Energy handed out generous grants to large-scale companies. By the late 1970s Boeing, McDonnell Douglas, Hamilton-Standard, Grumman Aerospace, General Electric, and Westinghouse retrained their space engineers and soon participated in the modification (MOD) wind turbine program. As might be expected, the economy of scale caught the fancy of engineers and the Department of Energy efficiency experts. Soon huge turbines began to sprout at Sandusky, Ohio, and Clayton, New Mexico; in Puerto Rico and Rhode Island; and at Boone, North Carolina, and Oahu, Hawaii. These so-called first-generation turbines varied in size from 100 kW to 5 MW. The program, funded at near $400 million, represented a much enhanced commitment for renewable energy. Did the industry reward the government's faith in wind energy? It did not. The problem was that the experimental turbines did not work, and the failures could not be hidden from public view. A large wind turbine must operate when the wind is up. If it is not spinning, public confidence is severely affected. The experiment already described at Medicine Bow, Wyoming, was fairly typical. When NASA's experimental 100 kW turbine in Sandusky, Ohio, failed in less than two days, the *New York Times* headline read: "$1 Million for Only 30 Hours of Work."[8] In defense of the program, engineers learned a great deal from failure. However, the politicians and the public were unsympathetic to that view. The last trial unit shut down in 1992, a victim of mechanical failures, vibration problems, and blade fatigue.[9] The Department of Energy's program reinforced the widely held belief that wind energy was an unworkable technology, and there was every reason to accept that interpretation. One megawatt turbine after another came to nothing, and these often catastrophic failures soon typified the disgraced face of wind energy.

The Enlarged Technology of Today

European companies, primarily Danish, settled for more modest expectations, building smaller, heavier, less sophisticated machines. But they worked! Today, large turbines conform to the dominant three-blade Danish style.[10] They have steadily grown in size and capacity, covering the hills of California's Altamont, Techachapi, and San Gorgonio passes. In the 1990s manufacturers began to experiment with so-called second-generation turbines, such as the WindMaster 300 kW, a Belgian-made machine, and Micon (600 kW), Vestas (600 kW), Bonus, Nordtank, Danwin, and Wincon, all of Danish origin. By the millennium century, the 1 MW turbine made an appearance. Many believed that the optimum turbine size had been reached. At a 1999 wind energy conference in Bellagio, Italy (see chapter 6), the majority of participants agreed that because of aesthetic considerations and maintenance and repair costs, the 1 MW turbine had reached the economy of scale.[11] We were wrong.

In 2008 the common size for commercial turbines is 1.5 MW to 3.6 MW. The accompanying chart gives an idea of a few of the more popular turbines:

Brand	Capacity (MW)	Tower height (ft.)	Blade height (ft.)	Total weight (tons)	Blade sweep (acres)
GE	1.5	212	338	164	1 acre
Vestas V90	1.8	262	410	267	1.5
Gamesa G87	2.0	256	399	334	1.5[12]
GE	3.6	—	40 stories	—	91,439 sq. ft.[13]

There are certainly other wind turbines available, such as the American-manufactured Clipper Liberty 2.5 MW turbine, Siemans (2.3 MW), Suzlon (2.1 MW), and Mitsubishi (2.4 MW), all considered fourth-generation turbines. What is most evident is that each generation of wind turbines increases in size and capacity. Recently Gamesa revealed that their turbine plant at Ellensburg, Pennsylvania, would be expanded. Gamesa spokesman Michael Peck admitted that "everyone's moving to bigger turbines. . . . We're moving to models that require a lot bigger workspace." In Germany the Enercon Company erected a 6 MW prototype, which surely stretches our concept of the economy of scale. With its immense blade sweep, such a huge turbine will operate in lighter winds, but whether that advantage is cost effective remains to be seen.[14]

One must ask if there is a place for a smaller turbine, one more in line with sensibilities like those of Joanna Lake. Outraged by the loss of some of Vermont's pleasing ridge lines by what she calls "monstrous turbine assemblies," Lake posed a question that always haunts Americans unconvinced by the obsession for continual growth. "What ever happened to 'small is beautiful'? Vermont is a small state." Reminiscent of disciples of E. F. Schumacher, Lake suggested that smaller windmills, generously subsidized, could offer a better solution.[15] Another Vermonter, Bill Roorbach, questioned the plan for forty-four turbines atop Kibby Mountain. Although sympathetic to wind energy, he reflected on whether the state's effort to conserve 4 percent of energy use (the amount the turbines would create) could not be achieved in a less obtrusive fashion. "Big wind" and big turbines disturbed his equilibrium. Joanna Lake wondered who would profit from the incessant drive for "noisy brightly lit monsters."[16]

The Possibility of Choice

It is hard to find anyone opposed to the concept of wind energy. However, many people question the direction of the industry. Is bigger better for the environment, or just for the owners of wind farms? Are we sacrificing the American landscape for private enterprise and the profit motive, rather than for environmental sustainability? The economy of scale, so important to the wind energy business, may have met the law of diminishing returns. I do not mean financial returns but, rather, environmental ones. People throughout Maine and elsewhere in the nation are questioning whether the electricity that the landscape-altering turbines produce is not outweighed by their transforming effect on the countryside. Observers often do not know how to react to the exchange of natural vistas for the titanic technology they see before them, so out of character with their memory or expectations.

Are there any options? Of course. First, local government entities and environmental groups need not allow engineers and corporate boards to make all the decisions. Developers are driven by certain principles, mainly the dictum that bigger is better, in both size and profit. Perhaps there is also a spirit of competition at work: who can build the largest wind turbine, whether is it cost effective or not? One is reminded of the international competition to build a supersonic passenger aircraft. The United States blessedly dropped out of the competition. The French-British consortium won, and now the impractical and environmentally destructive Concorde is part of the technological scrap heap of history. But that example has not influenced wind turbine manufacturers. At the moment wind companies approach a potential area with the philosophy that one size fits all. Wind developers, to my knowledge, never ask local leaders and affected residences questions like "What size and siting pattern would best fit into your landscape and your community?" or "What design alternative would best work for you?" Nor do wind developers suggest a design alternative that would integrate their product into the natural beauty or the historic use of a site. How could they? They rarely have more than one product available, and that product is immense. Ten years ago the wind companies created a good business with 1 MW turbines. If a community objects to 400-foot-tall blades, why not offer an option of 200 feet? Perhaps the wind farm would produce only 50 percent of the optimum, but acknowledging and addressing people's concerns would be a good investment. The important point is that impacted communities should have *choices*. Wind farm developers need to be sympathetic neighbors, and flexibility in turbine size would help accomplish that goal.

Can We Have Style?

I would like to see choices not only in size but in style. Today, there is one turbine style available, and the differences between manufacturers' turbines are hardly distinguishable. They are reminiscent of the one-year style changes of an American automobile. Obviously we cannot chose our windmills as freely as we choose styles of automobiles or clothes, ignoring key issues of reliability and efficiency. Yet aesthetics can play a role. For instance, in the 1980s the vertical-axis FloWind turbine made an appearance at Altamont and Tehachapi wind development areas. Although the vertical-axis machines offered a more pleasing profile, they were neither as productive nor as reliable as their Danish counterparts. FloWind eventually went bankrupt, but perhaps the design is worth pursuing or even resurrecting.[17]

Manufacturers have abandoned the eggbeater-style turbine, yet these attractive FloWind turbines at Altamont Pass may yet find a place in the wind farms of tomorrow. (Author photo.)

Wind energy expert Paul Gipe states that "inventors have built wind turbines in nearly every conceivable configuration, including Madaras's spinning cylinders, Schneider's lift translator, and various 'squirrel cage' rotors. Yet as wind technology has matured, only a few approaches have gained prominence."[18] Among all the rejected designs, Gipe does feel that the vertical axis merits attention. The Darrieus design commonly known as the "eggbeater" has advantages as well as limitations. My objective is not to compare the efficiency of the two styles but simply to suggest that manufacturers should make both styles available. Perception of wind turbines is highly subjective. I photographed a half dozen vertical-axis turbines (FloWind) in Altamont Pass in the late 1980s. One of the photographs appeared in my book *Wind Energy in America* with the caption: "FloWind vertical-axis turbines can be aesthetically pleasing. When the author took this photo, these units were turning in unison, as though performing a kind of ballet with the wind."[19] For me the Altamont windmill landscape, so vilified by many, was enhanced by the vertical turbines. They would be my design choice, and I would sacrifice a little efficiency, if necessary. When a team of four Americans and six Europeans met during the Bellagio conference to discuss the NIMBY response to wind energy, the group broke out evenly over whether the three-bladed Danish style should be considered standard. I and others felt there could be very different designs, as yet undeveloped.[20] I still believe that. What we must acknowledge is the creative power of young minds. The three-bladed turbine so prevalent today will one day be a first awkward try. I am convinced that humans will depend on the wind for power in 2111, but I do not envision that the generating devices will look like today's turbine. We have built huge, costly machines that have some environmental drawbacks, and they do not harvest the best winds.

Fortunately, the growth and potential of wind energy has spawned new ideas, making it abundantly clear that the windmills of today will wind up as historical curiosities tomorrow. What designs will emerge triumphant I cannot say, but engineers at FloDesign are working on a "jet pack" design expected will harvest three to four times more power than traditional designs. It will be easy on the eye, harmless to birds, and much less intrusive on the landscape. Other companies are devoting efforts to improved vertical-axis turbines, variations on FloWind designs.

The most revolutionary design would be unconstrained by a fixed tower and rotor. Meteorologists know that the strongest winds are far higher than our ability to reach them. The greater the height, the greater the wind, and the greater the power potential. But how are we to reach turbulence 2,000 feet up in the atmosphere? The engineering team at Makani Power expects to harness this high-altitude power, producing wind energy at below the cost of a coal-fired power plant.[21] Can this be done? Australian Saul Griffith, an inventive engineer educated at the Massachusetts Institute of Technology, has assembled a team of a dozen dedicated young engineers who are hard at work in the San Francisco Bay Area. The technology is rather difficult to visualize, but as Griffith describes it, "you'll just have a bunch of very large kites flying in circles all day, two thousand feet above the ground," tethered to earth with a thin transmission cable.[22]

Perhaps the vision of Makani Power will prove unsuccessful. If so, there are other designs in the pipeline. Young minds with fresh ideas are addressing the problem. Inventors and engineers worldwide will eventually, I believe, create inexpensive and novel devices to harness the immense potential of wind. It will only be a matter of time, and the time needed will be determined by the commitment of venture capitalists, ingenious engineers, and a sympathetic government. Above all, we need to have a choice. As communities across the country fight over whether to allow turbines on treasured landscapes, wind developers will do better to offer options and not simply dig in their heels. It is possible for people with opposing views to work out compromises, and the chances of this are magnified if there are options.

Texas Big

Wind turbines have grown steadily in size over the past thirty years, but this is only half of the equation. Wind energy developers need significant amounts of terrain as well as reliable turbines. I do not know when or where clusters of wind turbines became designated as wind farms. In the 1980s it seemed an appropriate descriptive term. Today it is more problematic. Wind farms have grown in acreage to the point of having much more in common with industry than with farming. We no longer measure their output in kilowatt-hours but rather in megawatts. Far from being just another aspect of a family farm, now they constitute a different actuality as industrial landscapes: power stations covering hundreds of acres.

The turbine-impacted countryside is often as expansive as the land will allow. Nowhere is this more evident than in Texas, which has plenty of space and abundant wind. The Lone Star State offers an excellent case study of wind plant growth, an example for the nation. In the fall of 2007, a surprising event occurred. The state of Texas surpassed California in wind energy capacity with 2,768 megawatts, while California rated second at 2,361. Some two years later the Texas growth spurt continued. Out in West Texas a resource that ranchers and oil and gas operators regularly cursed has become something of value. In the area of Midland and Odessa oil has long propelled the West Texas economy, but now wind energy had muscled its way onto the open landscape.

Of course the wind and the open space were always there, but it took willing investors and a sympathetic state government to unite and then transform much of West Texas. The economic catalyst is found in Senate Bill 7, passed by the Texas legislature in 1999. An extremely complicated utility bill, this Texas Electric Restructuring Act provided the basis for deregulation of electrical power in the state. Searching the eighty-page bill, I found one page devoted to renewable energy. That page set up what is called a "Renewable Energy Portfolio Standard (RPS)." Today approximately twenty-five states possess these portfolio standards. They mandate that a certain percentage of the state's generated electrical power must come from renewable sources, such as wind. The Texas RPS stated that the "renewable capacity in this state shall total 1,280 megawatts by January 1, 2003, 1,730 megawatts by January 1, 2005, 2,280 megawatts by January 1, 2007, and 2,880 megawatts by January 1, 2009."[23] Wind developers have easily exceeded each of these goals.

It may come as a surprise that George W. Bush was in the thick of Texas wind energy growth while he was governor. According to Pat Hall, who once headed the Federal Energy Regulatory Commission, Bush suggested that Hall take a new direction. After a 1996 meeting, Bush pulled Hall aside and said, "Pat, we like wind." Pat replied, "We what?" Bush replied, "You heard me. Go get smart on wind."[24] Was this an epiphany? Can we consider Bush an environmentalist who understood the importance of alternative energy? More likely he recognized a business opportunity for his state. Hall followed the admonition, and he and the commissioners found that many financiers were eager to invest in wind energy. Furthermore, many Texas consumers were willing to pay a premium for clean electricity. With the encouragement of the governor, the public utilities commission, and a sympathetic legislature, the Texas RPS detailed above became law.

Although the first commercial wind farm in Texas went on line in 1995 in the Delaware Mountains (35 megawatts), the real push came in 2001. In that year over 900 megawatts of capacity went on line in six locales:

King Mountain Wind Ranch (Florida Power and Light),
    278 megawatts
Woodward Mountain Project, Pecos County (Florida Power and
    Light), 160 megawatts
Trent Wind Farm, Taylor County, 150 megawatts
Indian Mesa, Pecos County (National Wind Power), 82.5 megawatts
Desert Sky Wind Farm, Pecos County (American Electric Power),
    160.5 megawatts

There was a lull in 2002–2003 when Congress failed to renew the federal tax subsidy of 1.9 cents per kWh produced (now 2.1 cents). This tax credit is for new construction and attaches to each turbine's production for ten years. But by 2005 incentives were again in place and new wind farms came on line. The largest in Texas and the world is Florida Power and Light's Horse Hollow Wind Energy Center. Completed in 2006, this vast wind farm has well over four hundred turbines spinning, with a capacity of more than 735 megawatts. The closest city is Sweetwater, which now boasts of being the wind energy capital of the world. Ironically, Sweetwater was the setting for Dorothy Scarborough's 1925 West Texas novel *The Wind*, a story of insanity and suicide as a result of the insistent wind and ferocious dust storms.[25] Times change; the wind now seems an economic benefit.

Keeping up with wind energy growth in Texas is like trying to describe a backyard squirrel's maneuvers through a pecan tree. Describe one development, and the turbines are off to another location. At this writing the American Wind Energy Association lists Texas wind energy capacity at 7,116 MW. The figure will grow. When recently asked about wind energy, Texas Governor Rick Perry responded: "Texas is wide open for business."[26] West Texas has plenty of land and plenty of wind, and wealthy entrepreneurs such as oil man T. Boone Pickens are intent on developing it. Pickens has put in an order for $2 billion worth of wind turbines, promising that by 2014 he will invest $10 to $12 billion to install 4,000 megawatts of wind power, although the sagging economy may slow him down. With hundreds of wind turbines, he hopes to revitalize the Great Plains with turbines spread all the way to the Canadian border.[27]

Pecos County, Texas

Doug May understands Boone Pickens's enthusiasm and fully supports it. May heads up the economic development corporation of Fort Stockton, a dusty little town in one of the driest regions of West Texas, surrounded by the vast expanse of Pecos County. Gone are the wild frontier days when Judge Roy Bean, the "Hanging Judge," dispensed his summary form of justice from the local saloon. Today the county prospers from subsurface resources: oil and natural gas. Although the oil reserves are diminishing, new natural gas wells are picking up the slack. So is the abundant wind. May knows that the natural gas reserves will eventually play out, but the wind will not. Wind energy represents a backup for those days when gas production shrinks. At present Pecos County boasts 474 turbines in play, mainly stretched out on the mesas that line the road to McCamey. They represent 403 megawatts of capacity.[28]

Pecos County encourages wind developers through generous property tax incentives. New development can expect a 100 percent county property tax abatement for the first five years of operation, although 10 percent of the figure must be donated to support a regional technical training center. Some of the center's graduates will eventually provide a maintenance staff for the numerous turbines. Of course, this abatement can be renewed or renegotiated. The number of permanent employees for the wind farms operating in the county is about thirty-five. This figure seems small, but Doug May likes to put it in perspective. The overall impact of employment in wind energy in Pecos County, population 16,809, is comparable to that of the Toyota plant in Bexar County (San Antonio area), population 1,417,501.[29]

One of the important payoffs for development of any new industry is in education. In most cases school taxes cannot be avoided by the companies, although in some counties the taxes are offset by capital construction contributions. In Pecos County the two school districts have been quite dependent on oil and gas revenues, which have fluctuated with availability and price. Now the addition of wind energy properties provides a more stable source of revenue. In such a thinly populated county the school tax revenue of $4,050,262 paid by the wind energy companies in 2004 has made a remarkable difference. Obviously, many residents are proud that wind energy is so environmentally advantageous, but lower property taxes add significantly to its popularity.

For the county, there are other benefits that would normally escape our attention. When I interviewed Doug May, he emphasized the symbiotic relationship of wind development and the oil and gas industry, particularly for the stability of the community. We know that both oil and natural gas wells will eventually play out and the workers will lose their jobs. However, May is convinced that these men and women will find work in the wind energy fields as welders, mechanics, electricians, and in all the jobs associated with heavy industry. May was proud to say that members of the Pecos County labor force would not have to pack their bags when the natural gas reserves are depleted.[30]

As we look at the Texas-sized benefits of wind energy, we ought not to forget the royalty payments given to landowners across the country. In northern West Texas, I accompanied Ken Starcher, director of the Alternative Energy Institute at West Texas A&M University in the town of Canyon, on a short field trip; Ken is a protégé of wind energy pioneer and physicist Vaughn Nelson. I met Ken twenty years ago when his beard was black. Today it is gray, perhaps from wind energy wisdom, for Ken understands the politics and intricacies of wind energy as well as just about anyone. The field trip took us to the White Deer wind farm and the ranch home of Robert and Marjorie Bichsel, who live in the shadow of about twenty large wind turbines. The Bichsels collect royalties from each turbine, are not especially focused on the addition to their income, and do not consider themselves to be making an environmental sacrifice. I remember my surprise when Marjorie Bichsel practically shouted: "I love these wind turbines." They genuinely enjoy living among tens of Mitsubishi 1-megawatt turbines, and for them the spinning movement seems to offer good company, and the noise was imperceptible when we were chatting inside their home.[31] Other landowners nationwide, however, are much more caught up in the economic possibilities, willing to endure any environmental fuss. Some have contracts that pay $2,000 to $8,000 in royalties per turbine per year. In some cases livestock or growing grain has become a secondary activity. It is the turning blades that ensure a livelihood and perhaps a few luxuries previously beyond a ranching family's expectations. Little wonder that Abilene area ranchers, always on the lookout for ways to survive, often seek developers to build turbines on their land, hoping to cash in on part of this new bonanza.

Opposition

On the morning of February 12, 2007, I drove north from Dallas to Gainesville and then west along the Red River to the tiny town of Saint Jo. I wanted to get a different view of wind energy in Texas. In Saint Jo I met Jack Shoppa, a land appraiser, realtor, and active member of the North Texas Wind Resistance Alliance (NTWRA). Jack wanted to take me for a tour, but a driving rain storm discouraged us. In his office he pulled out maps of a proposed Florida Power and Light (FPL) installation on bench land above the Red River that the utility had leased from largely absentee landowners. Jack made it clear that these landowners cared little for the community and were simply out to make money. If they talked green, it was the green of the dollar. Between the bench and the river spread about two miles of attractive riparian bottomland that had tripled in value over the past five years. With its accessibility to the Dallas–Fort Worth metroplex, this rural setting now attracts urban dwellers wanting a weekend retreat or country retirement home. Approximately seventy-five commercial-sized wind turbines would surely compromise the rural character of the region known as the Western Cross Timbers. Not surprisingly, a group of residents and landowners organized to oppose the project. Although their objections stemmed from an expected drop in land values if turbines went up, the NTWRA makes a case that wind energy does not make economic sense, and without the state and federal subsidies, it would never happen. Led by Joe Dial, NTWRA sued. FLP bowled over the limited forces of the NTWRA, and the case never went to court. The turbines are now up, but it is too soon to know if they have affected property values. Furthermore, in the present real estate market, proving that the turbines devalue adjacent land would be difficult indeed.

Thus not everyone in the state of Texas favors the development of wind energy. As a general rule, as one moves east, opposition to the wind stations escalates. Although no one has done a study, the most obvious reason is that central and northern Texas receive more rain and support a greater population, and in the eyes of many, the landscape becomes more desirable than in the drier west. As a general rule, this land dichotomy principle may apply to the nation as well, with opposition to wind turbines more pronounced along the heavily populated East Coast. The conflict centers around the highest and best use of land. Should desert and marginal ranching land be used for wind energy development, reserving the more vernal, hospitable, populated land for intensive agriculture and/or recreational use? Whatever the principal objection, all opponents find the turbines' remarkable size unwelcome.

A Failure at Rapprochement

For a model of organization, the NTWRA looked to a small group of landowners who were involved in their own fight nearby. About twenty miles south of Abilene (close by Texas standards) is the huge Horse Hollow Wind Farm, and right in the middle of it is Dale Rankin's horse ranch. Dale is an outspoken opponent of the hundreds of turning windmills with which he must now live. I called to ask if I might visit, and he said, "Sure." Driving to tiny Tuscola one enters a surrealistic world of movement; the turbines are so large as to be almost indescribable. In the distance I spied an eighteen-wheeler carrying a new turbine blade, and it looked like a toy Tonka truck as it paused at the base of a tubular tower. As I have noted, how people react to such a landscape is subjective. Some are entranced by the tall towers with 120-foot-long blades extended, sometimes turning in unison with adjacent turbines, sometimes not. The comparison that sprang to mind was Dorothy being swept away from Kansas to the Land of Oz, another world, a place of awe; and in the eyes of Rankin, a place of smoke and mirrors.

I was visiting. Living among the turbines is another matter. I turned off the county road and through Rankin's gate. The two-track led through a wooded meadow to a barn near a pleasant ranch house surrounded by a spacious lawn. Rankin was working with a couple of horses. He greeted me and we drove to the house. Inside his home I looked out through the picture window and instantly realized the impact of the wind turbines on his situation. Across a divide were two wooded hills, but the valley between them was occupied by the spinning blades of two mechanical intruders. I do not claim objectivity, but the turbines seemed inappropriate for the bucolic scene. For the Rankins the change is a sad story of landscape loss, evoking a need and passion for justice—and that is what Dale Rankin seeks.

I had brought my tape recorder, and Dale and his wife, Stephanie, were soon immersed in wind energy talk. I asked whether FPL representatives had approached them before siting the two offensive turbines. It seemed reasonable that a wise company executive could have mitigated or eliminated the turbines, considering that the project is a huge operation of some 421 turbines spread over 47,000 acres. Dale's answer was no. The Rankins did not own the land, and the utility company placed the turbines where its grid pattern determined they should be. Perhaps such a policy represented efficiency and good engineering, but such arrogance and poor public relations would force FPL into a lawsuit it neither wanted nor needed.

There are flaws with wind energy development in Texas, and the Rankin example explains some of them. The state of Texas does not regulate the siting of wind energy projects. A developer, such as FPL or Airtricity, simply approaches the county commissioners for a permit. Hardly any of the rural counties have planning commissions, so that hurdle is often eliminated. Most counties have only three commissioners, so finessing that step can be easy. Because these counties desperately seek jobs and development, a tax abatement for at least five years is almost automatic.

With few forms to fill out and few steps to clear, the wind company has its building permit. Of course, before the developer even approaches the county government, signed sealed agreements with landowners are already in hand. Landowners are becoming more sophisticated now, but the early contracts were one sided. Once the two parties agreed and signed a lease, often the landowner was sworn to secrecy, forbidden to discuss the details of the lease with neighbors, an attorney, or anyone else—a clear case of divide and conquer. The reality is that to realize and surpass its renewable energy mandate, the state streamlined the regulation procedure. As Governor Perry has remarked: "Texas is wide open for business."[32]

The only recourse for Dale Rankin and his eight neighbors was the court system. They sued in the 42nd District Court, claiming the turbines were a nuisance and that FPL was receiving excessive federal and state tax credits as well as local county tax abatements. By the time the suit entered the court in Abilene, the turbines were up, and Rankin's lawsuit was simply to stop further construction and seek compensation for the damage already done. In December 2006 a jury heard testimony on both sides, but Rankin's case was severely injured when the judge narrowed the legal complaints to only one: noise pollution. Visual intrusion, the judge ruled, could not be considered.[33] Given such limitations, the jury found that the turbines were not a nuisance to neighboring land owners. Trey Cox, FLP's attorney, summed things up this way: "Texas is very much a landowner's right state. We don't want neighbors fussing over what things look like. . . . As long as you're not doing anything illegal, if you want to have a broken-down barn or paint your house pink, you get to do it."[34] The decision was a disappointment to Rankin and the group's attorney, Steve Thompson. They planned to appeal.

In South Texas another bitter controversy drew attention. Two ranches of historic importance were at each other's throats over wind energy. One hundred and fifty years ago Richard King and Mifflin Kenedy, the founders of the King and Kenedy ranches, were great friends and "were closely associated in many ventures, and . . . both would rise to prominence in the frontier of South Texas."[35] No more. Jack Hunt, the president of the King Ranch, vehemently opposed the plans of the John G. Kenedy Trust to erect more than two hundred turbines close by. He also questioned the state's plans to lease offshore land for wind energy development. Jack Hunt based his opposition on destruction of scenery, disruption of bird migrations, lax oversight in the permitting processes, and excessive subsidies.

Both Hunt and Joe Dial of the North Texas Wind Resistance Alliance were vehement in their argument that wind energy does not make sense if the generous federal subsidies and incentives are stripped away. I describe such enticements at greater length in chapter 5; the important point here is that Jack Hunt's argument that the federal support paid both to the Kenedy Ranch and to the state of Texas (for offshore development) was excessive. In regard to offshore leasing, Hunt contended that the Texas Permanent School Fund would not be enhanced.[36] All in all, neither project made any economic sense, and they provided a perfect example of the waste of taxpayer money. In response, Jerry Patterson, commissioner of the Texas General Land Office, disputed Hunt's figures and noted that "despite Mr. Hunt's protest, all forms of energy are subsidized in the United States." To drive home his point, Patterson cunningly observed that "King Ranch Inc. has received millions in tax breaks on oil and gas production through such vehicles as depletion allowances, severance tax relief, sales tax exemptions and marginal [oil] well tax credits. Yet Mr. Hunt is now indignant about tax incentives for wind power."[37]

This imbroglio went on, and Hunt and his attorneys erected numerous legal roadblocks. However, U.S. District Judge Lee Yeakel sorted it out, and the Kenedy clan has won round one.[38] In January 2009 John Calaway, chief development officer for Australia-based Babcock and Brown, proudly helped dedicate the 283 MW Gulf Wind project, noting that it had overcome the objections of the King Ranch and that the development would "set an example for environmental stewardship." Babcock and Brown promises another 200 MW addition, which could make the project one of the largest in the world. Jerry Patterson was also in attendance, contesting the charge that he had violated the Coastal Zone Management Act. Sued a year later, in Texas style he stated that he had been called out "as a liar, and I don't cotton to that."[39]

In spite of occasional fierce opposition, Texas wind energy is growing and will continue to do so. The primary limitation is not turbines or siting controversies but power transmission. West Texas needs help to get its product to urban markets, such as Dallas–Fort Worth. Most of the grid in the state is under the control of the Electric Reliability Council of Texas (ERCOT). At times the grid cannot handle the electrical harvest of a windy day when the turbines are working at capacity. ERCOT figures indicated that occasionally only a little over 16 percent of the rated capacity of wind turbines enters the grid, and only about 3 percent of that was available during peak demand hours. Often the turbines were busy during the night but not turning during those hot summer afternoons of July and August, the time when Texans need the energy.

How much wind energy can the present Texas grid handle? Some engineers believe that only 15 percent wind energy is possible, while others say 30 percent. Paul Sadler, executive director of the Wind Coalition advocacy group, has said that "integration of wind is not sending a man to the moon. It's just a matter of having the will to do it."[40] Whoever is correct, all experts agree that the state must have more transmission lines. West Texas wants to produce electricity, and the major urban centers—Houston, San Antonio, and Dallas–Fort Worth—will consume it. But the state (or private investors) must transport the power from the place of production to the place of consumption. It sounds easy, but it is expensive. ERCOT has established what it calls Competitive Renewable Energy Zones, based on reliable wind regions and existing transmission line corridors. If private funding is unavailable, ERCOT is prepared to spend up to $10 billion in new grid construction with the intent of getting more wind energy onto the grid from these zones. Also, in July 2008 the Public Utility Commission of Texas voted 2 to 1 to authorize spending $4.93 billion to build new transmission lines from windy West Texas to populated East Texas. The state is already the nation's leader, and the new transmission lines will more than double the state's wind energy capacity to around 15,000 MW, a figure that would lead not only the United States in wind electricity but the nations of the world (see chapter 4). All of this is somewhat speculative. Bill Bojorquez, ERCOT's director of system planning, warns that wind energy can only supplement rather than supplant traditional forms of production, such as natural gas and coal.[41] Furthermore, the cost of this massive project will have to fall on electricity customers with a price rise of at least $3 to $4 per month. With the downturn in the economy and the already steep cost of electricity in Texas, such a plan is bound to face renewed opposition.[42]

Cape Wind

While Texas generates the most electricity from wind, Massachusetts has generated the most controversy. As mentioned in the opening paragraph of this chapter, in late 2001, Jim Gordon and his Energy Management Incorporated (EMI) proposed a project that would introduce significant wind energy production to the East Coast.

EMI seeks permits to put 130 turbines of 3.6 MW capacity on a stretch of water in Nantucket Sound known as Horseshoe Shoal. There the water is shallow, but the wind is strong. An offshore wind plant seemed ideal, especially in New England where, in contrast to West Texas, people are numerous and land is dear. If the eastern United States is ever to produce significant renewable energy, this seems the place. It has not happened. Seven years after Gordon's proposal, the wind continues to blow across the shoals uncaptured. The people of Boston, the surrounding towns, and out on the Cape rely on oil and natural gas for their power.

Opponents have raised many issues (bird deaths, fishing rights, boat collisions, etc.), most of which can be mitigated to varying degrees. The main issue is visual, involving the seascape view of the wealthy class centered around Hyannisport. This is less susceptible to mitigation. These people do not want to be looking at wind turbines on Nantucket Sound. In debate regarding the proposed turbines, John Passacantando, executive director of Greenpeace, warned the audience that they ought not to be distracted: the opposition to Cape Wind was based on the preference of the wealthy, and the rest was again smoke and mirrors. The title of a recent book, *Cape Wind: Money, Celebrity, Class, Politics, and the Battle for Our Energy Future on Nantucket Sound,* so accurately describes the issue that one might skip the contents.[43] Yet the details are engaging. Old money families (Mellon, DuPont, Kennedy) have combined with new wealth residents (Jack Welch of General Electric, Paul Fireman of Reebok, Douglas Yearley of Phelps Dodge) and vehemently opposed this renewable energy development.

This, of course, is a classic case of not in my backyard. The wealthy of Cape Cod do not care how much they pay per kilowatt-hour for electricity. It does not matter. They will fight passionately, however, for their view from the picture window and for their sailing rights, even as they claim to care about global warming and our deteriorating environment. They are highly educated and fully aware of the consequences of our profligate use of energy, in which they are notable participants. The late Senator Ted Kennedy, for example, chastised the Republican administration for opposition to the Kyoto Treaty on global warming. He opposed oil drilling in the Arctic National Wildlife Refuge. Yet he and his nephew, Robert Kennedy, Jr., have been at the forefront of opposition to the Cape Wind project. A NIMBY response is seldom based on science, fact, logic, or environmental concern; it usually derives from an emotional response to place, particularly change to a treasured landscape or seascape. Contradictions abound in the after-sailing cocktail conversation at the exclusive Hyannisport Yacht Club.

Part of the problem is the complex jurisdiction on Nantucket Sound. Strangely, the U.S. Minerals Management Service, an agency at home in a different setting, is overseeing the review process. With such a controversial topic, review has been rigorous and comprehensive. The Draft Environmental Impact Statement contained 5,000 comments and is 3,800 pages long. Bureaucrats have sifted through this almost unprecedented number of arguments, rebuttals, and rejoinders.[44] Given the wealth, power, and determination of the opposition, a well-paid and determined staff of attorneys can use the court system and the permitting process to create numerous roadblocks, much as in South Texas. This is their strategy. They bewail how the environmentalist's dream is their constituency's nightmare. Whether Gordon can withstand the long delays, negative publicity, and permit obstacles remains in question. Yet he is determined, and he responds to questions of uncertainty with: "Not *if* the permits are granted. *When* the permits are granted."[45]

In January 2009 Gordon received good news when the Final Environmental Impact Statement declared that the proposed Cape Wind project posed no serious environmental threat. In most cases such a positive report should seal the deal, but not with Cape Wind. The Alliance to Protect

Nantucket Sound executive director immediately issued a statement that "Cape Wind is far from a done deal, despite this favorable report." Senator Kennedy continued his opposition, stating that "by taking this action, the Interior Department has virtually assured years of continued public conflict and contentious litigation."[46] It is sad to see a viable project opposed by the forces of powerful politics and self-interested wealth. Hyannisport would leave a more honorable legacy by gracefully conceding that the fight was over instead of continuing to contrive barriers. However, no one should bet against Jim Gordon. The shift of consciousness at work in America may affect even the wealthy of Cape Cod, and Gordon and his company expect to be generating wind electricity by the end of 2011.

The wind energy business has changed dramatically in the last two decades. In 1990 wind energy had only a tiny impact on the nation's electrical needs. True, there were thousands of generators turning in California, but six of today's General Electric's 1.5 MW turbines produce the same amount of electricity as ninety of the first-generation U.S. Windpower turbines.[47] The wind farms of the 1980s and 1990s were little more than a band-aid for our oil habit. Now both the size of the turbines and the wind energy installations promise to make a difference. Yet there are always trade-offs. From 1980 to 2010 the turbines have grown in size and power, but we must remember the environmental axiom at work here: "For every gain there is a compensatory loss."

CHAPTER 3

The Riddle of Reliability

I feel like I am putting Humpty Dumpty back together again.
—Lawyer Lantson Eldred, on reliability at Cabezon wind park

"Reliability" is the most important word in the wind energy developer's lexicon. Why so? Because in the past wind turbines have been so unreliable. They could not be trusted to perform. The first generation of turbines was like a junker car that might or might not run when you turned the key. They were undependable, neither solid nor sound in construction. I have referred to Cabezon wind park in the pioneer days of California as a graveyard of downed turbines and twisted blades. In trying to resurrect the Cabezon clutter, lawyer and investor Lantson Eldred invoked the nursery rhyme about trying to put Humpty Dumpty back together again.[1] During this pioneer period of technological bedlam the only group that profited from the situation was the attorneys, who happily sued unreliable turbine manufacturers, untrustworthy developers, unresponsive government officials, and just about anyone else associated with the wind energy business.

When new technology emerges, we should not be surprised if failures outnumber successes and costs exceed income. Building a reliable wind turbine proved difficult. After all, a turbine must harvest a capricious energy source. Imagine the strain on machinery that must constantly adjust to winds ranging from 8 to 60 mph. To put this into perspective, consider the life expectancy of an automobile's engine if one were constantly shifting the revolutions per minute from 1,000 to 6,000 and all numbers in between. Furthermore, wind turbines are unprotected, exposed to sun, wind, rain, snow, and in some areas, blowing sand. Engineering a reliable wind turbine took time, talent, and perhaps even a little luck.

Part of the reliability problem is the variable nature of the wind itself. Is it possible to make a variable resource reliable? Can we force the wind to do what we want it to do? We cannot. But the latter part of this chapter covers scientists' efforts to *predict* wind strength, on not only a daily basis but hourly as well. If meteorologists can predict the variable wind patterns accurately, reliability of wind electricity can be more than doubled.

Keeping Them Turning

According to historian Terry Reynolds, the average life expectancy of a waterwheel during the colonial and early national period of the United States was ten years, although owners could extend this through proper maintenance.[2] We can probably assume that large wooden windmills would have had the same life expectancy. The friction of moving machinery lends itself to wear, especially when gears of wood are involved. Metal has the advantage. The simple American windmill, developed for the Great Plains, had largely metal parts and was self-lubricating. The working life of a self-oiling windmill was considerably more extended. Save a wind event such as a tornado, it could be expected to continue to pump for forty years. Although reliable, it did require attention, often performed by a cowboy saddled with a new job of climbing lattice towers and tending a large ranch's numerous windmills. If he tended toward vertigo, he might seek another occupation.[3]

Whereas the American water-pumping windmill was a relatively simple affair, the farm wind turbine was not. Based on aeronautics developed during the Great War, the propeller turned six to ten times faster than the vanes of the water pumpers in equal wind velocity. Owners also had to contend with a generator and a passel of storage batteries. Many people found electricity a mystery best left to professionals, but there were no electricians available on isolated ranches a hundred years ago. Wind turbines were a finicky novelty, consistent performance was rare, and anyone not handy with electricity dreaded working on a turbine 50 feet in the air. Most ranchers and farmers preferred a gasoline generator or just managed without electricity.

Survival in the wind energy business required building a trouble-free, unfailing machine. The Montana-born brothers Joe and Marcellus Jacobs accomplished this feat. As noted, reliability was the hallmark of their turbine. They managed to build a balanced turbine with the correct pitch of the propeller blades, a variable speed system, and a reliable generator. The Jacobs brothers sold this package, with a tower and battery thrown in, for about $1,000. During the depression years this was a fancy price, and yet the company sold some thirty thousand units, primarily because they represented quality. They worked, many of them for decades, without need of repair. Engineers have learned much about the mechanical peculiarities of turbines since the days of the Jacobs boys, but operating performance — that is, the production of electricity day in and day out — has been tricky.

Elusive Reliability

The huge commercial turbines that we see today have their origins with their Danish predecessors but also with the Smith-Putnam 1.5 MW machine erected on Grandpa's Knob in Vermont in 1942 (see chapter 1). The Smith-Putnam turbine performed superbly, considering that it was the first of its kind, was built from scratch, and employed all kinds of new technologies and untried ideas. It was wind energy's equivalent to getting a man on the moon. Although it would eventually fail from blade fatigue, it proved dependable for many months, and it resulted from the work of some of the best scientific and engineering minds in the country. But there was no follow-up, no continuation. Knowledge, no matter in what field, is a cumulative process. Soundness in design would result from Danish repetition, modification, and adjustment: the overlaying of one experiment (and turbine) with another to take wind energy knowledge to the next level. For reasons already examined, the S. Morgan Smith Company and Putnam could find no support, either private or public, for continued work.

Thus, when the federal government took an interest in renewable energy in 1974, engineers had no real baseline of knowledge from which to work. They could not leap into the future without some baby steps. For space engineers reassigned to work on wind energy, reliability had previously meant getting a rocket free of earth's gravity and into space in a few minutes — high-tech stuff. They had been spectacularly successful in conquering space with technology that had to work for a brief, exciting moment. Now they set to work to build a large American wind turbine, a production that would employ advanced technology and computers and yet would retain the qualities of a draft horse rather than a rocket. This was not romantic work, for what was needed was a turbine that would produce electricity hour after hour, day after day, night after night, for years. And if they succeeded, there would be no celebrations or ticker tape parades.

Were Americans up to the task? The answer was no. The engineers' learning curve had simply skipped over simplicity. I have already described the suite of failures of the late 1970s and the 1980s. Not only could we not build a reliable horizontal turbine; neither could we build a vertical one. The failure of the Alcoa vertical-axis 500 kW turbine is legendary. Placed on a San Gorgonio Pass site, and powered up with some fanfare, it lasted two and a half hours before a brake failure caused the rotor to accelerate and eventually resulted in a total collapse.[4]

The Danes Show the Way

What was the problem? One is reminded of Henry David Thoreau and his advice to Americans to "simplify, simplify, simplify." Engineers had a failing grade in reliability and must learn what *not* to do. They had a good model to show the way but not one they cared to follow. During the middle years of a turbulent century, the Danes developed simple turbines for a particular need. During World War II the Germans cut off the coal supply to the diminutive nation. After a cold winter and desperate for a little electricity, skilled craftsmen developed simple turbines to meet a need. It was a lesson in self-sufficiency that the Danes never forgot. More recently Matthias Heymann, a German scholar, made a rather remarkable discovery. Between 1973 and 1988 the United States spent $380.4 million on wind technology, and West Germany budgeted $78.7 million, while Denmark was a distant third with $14.6 million. The success rate revealed a reverse positioning—a modern parable of the tortoise and the hare. The two hares, the United States and Germany, ripped along with plenty of funding and publicity but ultimately failed. Diminutive Denmark followed a different path. Capitalizing on its historic war experience, and the basic knowledge provided by Poul la Cour and Johannes Juul, the Danes easily triumphed in the global competition for that elusive large, dependable turbine.[5]

A further explanation is in order. First, the Danish experience under-scores the importance of cumulative knowledge; those baby steps before you leap. No nation had a stronger tradition of use of the wind, nor were the Danes seduced with the promises of nuclear power. Second, they em-phasized what we might call a low-tech approach. Matthias Heymann calls it "the craftsman tradition."[6] Inventors lacked advanced degrees and were often practical amateurs who might work for a farm implement fac-tory. They used plenty of steel in their wind turbines, and American engi-neers often thought the results clunky, unsophisticated, heavy, and inef-ficient. So they were, but while American and other European turbines were frequently down for repairs, the Danish machines kept turning and turning.

The slow and steady performance of the Danish turbines clearly won the field in California. In 1987, 90 percent of new installations were Danish-built.[7] While American companies slipped quietly away, such Danish wind turbine companies as Micon, Bonus, Wincon, Danwin and Vestas prospered. From this California competition emerged what is often called the Danish design. It featured a three-bladed, upwind design (i.e., the front of the blades facing upwind from the tower), medium sized, with a fixed-pitch hub, stall control, and a lower rotor velocity. Again, the pri-mary themes were heavy-duty construction and technological simplicity.

Of the Danish companies, Vestas has emerged dominant today. The company's turbines may be found throughout the United States (and worldwide as well). For many years American did not know Vestas. Re-cently the company embarked on a high-profile advertising campaign that features the value of wind electricity as evidenced by their graceful tur-bines. Not only do their turbines dot the landscape, but the company has built tower and blade plants, bringing needed jobs to the nation. What BMW is to automobiles in the United States, Vestas is to wind turbines.

Initially, progress was not easy. Even Vestas had much to learn in the large-scale context in California. At the SeaWest Energy Corporation facil-ity, an early wind farm in San Gorgonio Pass, fatigue cracks appeared in fiberglass blades in over half of the Vestas units. The wind farm was forced to shut down while Vestas engineers designed more reliable blades. At Tehachapi Pass one field manager explained turbine failure from fierce winds that lodged small pebbles in the nacelle's machinery 150 to 200 feet high.[8] Unanticipated structural and environmental issues plagued the early wind energy business. The trajectory resembled the pioneer period of aviation when almost as many planes crashed as flew, and like airplanes, turbine failures could not be hidden from view.

Does a Warranty Guarantee Reliability?

The pioneer period of wind energy is now over. In the last decade wind turbines have evolved from a new, difficult, and often unsuccessful technology to one of much greater efficiency and reliability. This generalization applies as well to American companies such as General Electric and Clipper Windpower. The companies have learned from the Danes. For instance, with the exception of start-up and cut-off, the modern commercial turbine rotors turn at a constant speed no matter what the wind velocity. This was not always the case. On the old U.S. Windpower turbines at Altamont Pass the rotor accelerated at each gust of wind, only to slow down afterward, which was hard on the machinery. Today's commercial turbine rotor spins at the same speed because of the induction generators used. Their effect is similar to that of a "governor" on a truck or car, controlling the engine's revolutions per minute no matter how hard the driver steps on the accelerator. Thus, although there might be a substantial wind, the modern turbine appears to be turning slowly. It is, but because of its broad sweep, the tip speed of the blade is around 120 to 160 mph. Could the turbine create more electricity without the governor? Yes, but as Paul Gipe points out, "designers are willing to sacrifice some performance for the simplicity of fixed-pitch blades driving a constant-speed generator."[9] The modern design features simplicity, which translates into that elusive quality called reliability.

With advances in performance, turbine companies now offer a limited warranty. Sometimes this can be expensive. In 2007, when Vestas Wind Systems turbine blades began to fail on the Platte River Power Authority's Medicine Bow Wind Project, the company covered the expense of as much as $100,000 per turbine. The warranty expires in 2011, at which time the power authority will be responsible for repairs, according to manager John Bleem. Hence they are negotiating a service contract.[10] Realistically, the repair cost of a drive train and/or generator situated over 200 feet in the air can be expensive, difficult, and dangerous. If a crane must drop the nacelle (housing the drive train, generator, etc.) to the ground, the repairs can easily eat up the turbine's profits for a year. Forgoing a service contract could pay off, but it is a risk, perhaps comparable to canceling insurance on one's house.

What kind of warranty does a turbine manufacturer provide the purchaser? Perhaps again an automobile analogy would be useful. With my 2005 Subaru the company provided a three-year, 36,000-mile guarantee on parts and labor for any repairs. I used the warranty twice, once with success (an electrical switch) and once with failure (rear wheels incorrectly aligned). Sometimes warranties work, sometimes they do not. The cheap replacement part the dealership paid for; the expensive one (four new tires) came out of my pocket. So what about a wind turbine, a product that tops $2 million in price?

As an example, the warranty between manufacturer and user on General Electric's popular 1.5 MW turbine is negotiated and complicated. In 2001 GE/Enron provided a five-year "bumper-to-bumper" package on each turbine. However, to activate the warranty required a payment of $15,000 per year per turbine, and an operating and maintenance cost of $14,000 per turbine. In essence, the operator was not receiving a warranty so much as purchasing an insurance policy against any breakdown and consequent loss of power (cash flow). GE's more recent package is about $25,000 for the warranty and $25,000 for service, and there are higher rates. There is a certain irony in purchasing a mandatory service contract to validate a warranty.[11] My Subaru's 36,000-mile warranty has now expired. Insurance companies urge me (unsuccessfully) to buy an extended warranty. But at least the first 36,000 miles were free, a luxury wind energy developers do not enjoy.

Other manufacturers put other conditions on their turbines, but what is clear is that a warranty is a pledge resting upon the good faith of the manufacturer. Any warranty includes so much small print and so many loopholes that ultimately the wind turbine user is at the mercy of the reputation and business ethics of the turbine builder. Any serious breakdown under warranty calls for negotiations, often arbitration, and perhaps attorneys. With luck things can be worked out amiably.

In summary, wind energy manufacturers have made significant progress since the 1980s. It would be impossible today to purchase the kind of junk that littered the hills of Altamont Pass and San Gorgonio Pass. The new turbines represent sophisticated knowledge born of Danish traditions, honed with European experience and the hard knocks of American trial and error. Operators wish manufacturers could simply issue a guarantee for at least five years, but that is yet to come and may never arrive. Wind turbines are designed to last for twenty years, but at present most manufacturers and their clients do not care if they last that long. The value of the turbine in terms of tax benefits and accelerated depreciation is complete within ten years. After that period a turbine must pay for its keep strictly through electrical energy production.

The Benefits of Competition

One more aspect of reliability that should be mentioned is the benefits of competition. To return to the automobile comparison, meaningful warranties are the result of fierce competition. Imagine the plummeting sales of a car company that offered no quality guarantee, or one that car owners had to pay to activate. After World War II automobiles came without guarantees, and people were happy just to be able to purchase a car. But as competition increased, particularly from foreign cars, warranties became commonplace. At the moment there is little competition among turbine producers; the CEO of Germany's Enercon, makers of the turbine considered the most reliable on the market, announced that he would export no turbines until the United States was out of Iraq! If companies have back orders for at least two years and the likelihood of continued demand, why would a company guarantee a product as risky as a wind turbine? The time will come. As the turbines become perfected and supply exceeds demand, one day a wind turbine manufacturer will announce a comprehensive labor and parts guarantee for five years, or perhaps even longer. Other manufacturers will fall in line, and the day of the truly insured and reliable wind turbine will have arrived.

The Reliability of the Wind

The other puzzle of reliability is the wind itself, our primary energy source. We know many of the attributes of wind, but one definite negative is consistent performance. Those who denigrate the practicality of wind energy always harp on its capricious nature. Some days the wind is creating havoc at 50 miles an hour; other days it is at rest. Those other days cause engineers and utility people to go berserk. How can you possibly run a utility in which customers demand constant access to electricity but human beings cannot control the energy source? How can you manage an efficient grid system when one of the energy sources is so fickle? I discuss the electrical grid in some detail in the next chapter; suffice it to say here that the unstable nature of the wind represents the strongest limitation to its use. All engineers and wind energy advocates acknowledge that the wind's energy is unlikely ever to provide more than 25 percent of the nation's power needs because it is undependable. Any higher percentage will be a very expensive challenge. Unless, as a nation, we are prepared for frequent brownouts and blackouts, wind energy electricity will always have limitations.

Yet those engineers and grid gatekeepers who throw up their hands and object to dealing with wind energy are somewhat in denial. Population increases and insatiable demand for electricity can put the electrical grid system at risk no matter whether it is charged with wind energy or not. When New England went into a dangerous cold snap in 2004 the grid operator (ISO) in Holyoke, Massachusetts, called for backup natural gas plants to come on line. But there was no gas available. Plants could not come on line, and others had to go off line. Schools and businesses closed. The grid almost went down, but warmer weather averted a catastrophe.[12] The point is that nature's resource may not be reliable with regard to power generation, but neither are human resources. Crises happen, and blackouts will occur. Engineers responsible for unfailing electricity must learn to live with the exigencies of wind power, not dismiss them.

Predicting the Wind

One way utility companies are learning to manipulate wind energy is through expanded meteorological knowledge. Scientists now have accurate wind maps, spelling out the average wind speeds for each region of the country. And all companies do local onsite testing with anemometers, usually for a year, to determine average wind speed. Before investing many millions of dollars a company must have accurate knowledge of what to expect in the region where it will do business.

Wind prediction is a new science. Meteorologists' ability to foretell rainfall has advanced, and although predicting wind velocity is more difficult, those who practice this science are getting more and more accurate. Today there are companies that specialize in predicting wind, and they are hard at work, determined to minimize the intrinsic uncertainty about wind speed. Wind forecasting is one of the primary tasks at the 3TIER Company, for instance. The company seeks to assist wind farm operators in predicting short-range wind energy production because it considers "one of the greatest challenges to integrating large amounts of wind energy has to do with the inherent uncertainties of the wind."[13] By "integrating," the company's scientists seek to get the maximum amount of wind energy into the grid.

Another company, WindLogics, employs nearly forty meteorologists, atmospheric scientists, and computing experts to make wind assessments and predictions through modeled weather data. The company can predict day-ahead and hour-ahead wind speeds and, consequently, energy production.[14] Obviously, an accurate forecast assists a wind installation manager with operational and marketing decisions, and it can also provide vital information for those operators responsible for the energy mixture on the grid. As techniques improve and knowledge increases, it will be possible for grid managers to expand the percentage of wind energy that the various grid systems can accommodate.

Prediction will also help to alleviate some inefficiency. Occasionally coal plants and wind turbines are working at cross purposes, resulting in wasted electricity. Coal plants take as much as twenty-four hours to ramp down from peak production of electricity. Often when wind energy production is up, coal plants have not yet ramped down, resulting in an overlap of electrical production. Although the waste is often insignificant, it can be minimized if meteorologists can give an early, accurate forecast of when additional megawatts of wind energy will come on line. The secret for grid operators is balance. They must feed into the grid what is needed. If they know the available sources, balance is easy to maintain. Wind energy, with its production determined by nature, is the most difficult variable. If it is more predictable, then the grid system will be spared duplication.

A Skilled Workforce

Reliability is dependent not only on engineers and meteorologists but also on a competent work force to keep the turbines running. When wind energy was in its infancy, the windsmiths learned their trade on the job and by the seat of their pants. In California many workers came from Denmark, Germany, and Italy, lured by a steady paycheck as well as the excitement of participating in a novel experiment. Those heady days have been compared to the boom of an oil rush. Whether the comparison is apt is questionable, but clearly working on wind turbines was not your normal eight-to-five job. There was a certain sense of community and pride in achievement. Ty Cashman, who went came to California in 1979, recalled that he "and a lot of others believed a new world was possible. We had the confidence that we could do something really different—and do it quickly."[15] As we saw earlier, 1980s participant Paul Gipe compared it to a barn raising.[16]

Working on huge wind turbines can be exciting. Rappelling down a turbine tower, a windsmith must combine mechanical skill with the daring of a mountain climber. The pay is good and workers are afforded every protection, but it is not for the faint-of-heart. (Clipper Windpower–NREL photo.)

Today most workers are trained for the job. Often recruited from the unemployed who, until recently, were building houses, installing swimming pools, and working backhoes, they are retraining in a growth industry, where their skills are in demand and will be appreciated. At Cerro Coso Community College in Kern County, California, instructors retrain the displaced as repairmen for the mammoth turbines. An eight-week course costing $1,000 in tuition provides a new skill and more secure employment. But clearly the life of a windsmith is not for everyone. You have to have good knees, and a stomach for heights. Instructor Merritt Mays puts in plainly: "I've seen guys freeze halfway up the tower." One of his most compelling lectures to students is "How to Avoid Death and Dismemberment." Bob Ward, an employee of General Electric, states that the job is more complex than one might think, combining the skills of a top mechanic with the stamina of a mountain climber.[17]

James Madison University students install a meteorological tower in Quinby, Virginia. If the anemometer measures sufficient wind, the next task will be to put up a small wind turbine. Such educational activities may lead to future jobs, and students have the satisfaction of knowing they are doing good work for the environment. (Photo by Jonathan Miles, NREL.)

For a young person with a mechanical aptitude and a good head for heights, a work life high in the turbines offers a wonderful opportunity. Misty Eisenbarth, a young single mother of two, was earning $6 an hour at a fast food restaurant. She now works as a windsmith in Wyoming, climbing towers and maintaining turbines in a large wind farm. Her co-workers, mostly male, accept her, and she has an "awesome job" with good job security.[18] For workers like Misty, looking after thousands of new machines yearly and maintaining near thirty thousand already in operation, unemployment is the least of their fears. Doug May of Fort Stockton, Texas, believes these turbine maintenance jobs will be with us long after the oil and gas workers have captured all the fossil fuels from the land.[19] From the point of view of the industry, skilled and committed workers are essential. Like airplane mechanics, workers employed in this new trade understand that they represent a crucial link in the chain of responsibility and dependability.

Unit Size

Mechanical and timing questions aside, certain aspects of reliability do favor wind energy. As Philip Schewe has noted, "ever since the time of the grid's founding fathers, power plants had been getting bigger. This was the economy of scale." However, Schewe now believes that utilities, much like financial institutions, have exceeded the economy of scale and that bigger is no longer better. What is his logic? It has to do with reliability of a 1,000 MW coal-fired plant: "A failed part, even a small one, could mean that the whole 1,000 megawatts would have to be taken out of play." In the late 1960s Consolidated Edison's huge "Big Allis" generator was "chronically ill," as Schewe puts it. The result was that Con Ed had plenty of trouble meeting peak power demand during summer heat.[20]

If a major unit like Big Allis or a nuclear plant goes off line, the possibility of a systems failure is palpable. In our present age, so is the possibility of terrorist action against coal or nuclear plants. Although the chances of systems failure or terrorist attack are not high, a decided advantage of wind turbines is diversity. A 1,000 MW wind station equals four hundred 2.5 MW turbines spread over a number of square miles. A terrorist attack could not easily result in catastrophic failure; more likely, the power output of only 5 to 10 percent of the turbines would be down at one time. When Percy Thomas attempted to convince Congress in 1950 to appropriate monies for huge turbines, perhaps his most effective argument was the idea of energy diversity. If the enemy attacked vulnerable oil or coal plants, the wind turbines would remain free of sabotage.[21] Thomas did not convince Congress, yet widely dispersed, decentralized power plants do have a strategic advantage in an age of terrorism. The message is that if we want reliability in the electrical grid system in the United States, we should build in diversity.

As a final observation about wind energy and reliability, it is useful to take the long view. Will oil, coal, natural gas, or uranium be plentiful in one hundred years? Probably not, if the present is any predictor of the future. If we do have these resources available, we know they will be prohibitive in price. Presumably other energy sources must fill the gap. Whatever changes occur, the free wind will continue to blow as long as the earth and sun exist. That is reliability we can count on.

CHAPTER 4

Tying into the Grid

[The grid is] the most significant long-term barrier to continued wind power expansion.
— American Wind Energy Association

Almost everyone in the United States uses electricity, but few of us ever think about it. Only in its absence does electricity command our attention. Perhaps our electrical service should be disrupted for an hour once a week, just to remind us of our dependence and set us thinking about electricity and how it is produced. The electrical grid is, of course, the transportation system that moves our wind turbine electricity from the often distant generating site to our homes, where it gives us light and performs innumerable tasks, making our lives much easier. One hundred years ago electricity was optional. Today it is essential.

The grid's job of getting electricity from where it is produced to where it is consumed seems simple enough. That should be the end of the story for wind energy, but it is not. Wind energy has created unique problems. For commercial wind energy to thrive it needs grid capacity, for short distances or long. The American Wind Energy Association has identified transmission on the grid as "the most significant long-term barrier to continued wind power expansion."[1] It is an immense problem. In early 2008 the Electric Reliability Council of Texas (ERCOT) estimated that the cost of building enough transmission lines to handle the growth of wind energy in West Texas could be as high as $5.75 billion if all the wind projects now proposed, equaling 46,623 MW, were built.[2] That megawatt figure will not be met in the foreseeable future, but 24,000 megawatts is not unrealistic. Thus, although the technology can be mind boggling, we need to take a look at the electrical transmission system and how it affects wind energy and vice versa.

How Does It Work?

Technically the transmission system would extend into each electrical socket in your home. But because that final tentacle of the immense octopus is the job of the local utility company, we will leave that distribution for another time. Until the last thirty years the tasks of acquiring energy sources (oil, natural gas, coal, nuclear), generation, transmission, and distribution were the monopoly of the utility company. In the San Francisco Bay Area where I was raised, the Pacific Gas and Electric Company (PG&E) acquired the energy source (often hydroelectric), generated the power, transmitted it on their network of high-voltage lines, and then distributed it to our home. One company did it all, and whatever you consumed would be paid for with one check, which my father always thought was too high. We had the California Public Utility Commission to protect us, but my father (and many others) did not believe they were doing their job. Like our telephone, the company represented a monopoly, and unless you did without electricity or generated your own off the grid, you were at the mercy of that company. We now have deregulation, at least in Texas, which has brought competition but not necessarily lower rates.

But how does the grid work? Some months ago I asked that question of Pat Hall, a Texan who was head of the Federal Energy Regulatory Commission during the Reagan years and is presently an administrator and developer for Airtricity, an Ireland-based wind company operating in the United States. Pat mulled over the question, no doubt searching for a simple answer. Eventually he suggested I should think of the grid as a huge bathtub of electricity with new sources continually pouring into the tub. At the bottom, Wood explained, the electricity exits through the drain to be transmitted and distributed. I liked that explanation, although in time the bathtub analogy furrowed my brow. It implied that electricity could be stored, and even I knew that engineers have yet to devise a system to store large amounts of electricity. So no bathtub. We either use it or lose it.

Perhaps the metaphor of a river would work better for the grid—a significant body of electricity *moving*, with tributaries, large and small, adding to the stream. Some rivers flow into the sea, but when reaching its goal, this river of power is channeled and consumed, and then it will return. The "grid master" (in Texas this would be ERCOT) records the amount of each stream that flows into the grid, and the producer of that stream is paid accordingly.

The human body can offer an alternative explanation analogy. The heart is the power plant, pumping blood to the arteries and capillaries, to return once again to the heart. Returning is essential to the heart and also to the grid. The grid's circulatory system is a *circuit* in which electricity moves alternately back and forth (alternating current, AC), clockwise and counterclockwise, at speeds we can hardly comprehend. Electrons are always moving, much as the blood in our bodies does. The circuit "loads" (provides power or watts to) a light bulb or motor and then returns to the power plant. So the circuit makes a round trip to be recharged. Breaking the circuit at any point stops the current. As Brian Hayes points out, "The opposite of an open circuit is a short circuit; when the outbound and the inbound conductors touch, bypassing the load: Pffft!"[3]

In Dallas our electricity provider is Green Mountain Energy, which advertises itself as using 100 percent renewable sources, mainly wind with a little hydro. When I first switched to Green Mountain, I asked about this. When the woman who took my application said that the power source was 100 percent wind, I replied, "How can that be, for the wind does not blow all the time?" She became a little flustered, replying, "All I know is that it is 100 percent wind power." Of course the *actual* electricity our household consumes is in all probability produced from coal marketed by Texas Utilities (now called Oncor), the dominant energy provider. The small wind energy stream flowing into the grid is nevertheless measured, and Green Mountain is paid the per kilowatt rate that is in my contract. The more such contracts, the more capital Green Mountain will have to bring new turbines on line or purchase wind power from other producers.

The Growth of the Grid

The grid system in the United States is an incredibly complex network of spider webs, interconnected at most points but occasionally independent. There are eight regional grids operating under the umbrella of the North American Electric Reliability Corporation.[4] It did not start out that way.

When in the late 1880s Charles Brush began stringing arc lights from a single generator, the Cleveland inventor gave us the first grid. Thomas Edison soon followed with his Pearl Street generator and a line stretched out to illuminate the homes of ninety-three wealthy New Yorkers. Many such grids followed. But these were single-source grids that have little in common with those of today.

The story of the Giant Power proposal in Pennsylvania in the 1920s is a closer parallel. It was, perhaps, the first effort to pool (or stream) different power sources. Not only did it integrate coal, oil, and hydro sources, but it would involve both private investor-owned companies and public (often municipal) service agencies. I should say that the system failed to be built, but it is worth examining the grid plan for the knowledge it sheds on a significant social and political issue of the day: Should the power system be controlled by private hands or public regulators?[5]

Both the advocates of Giant Power were committed to publicly owned power. One was Gifford Pinchot, then governor of Pennsylvania but nationally known as a Progressive conservationist who headed the United States Forest Service during the administration of President Theodore Roosevelt. Pinchot intensely distrusted and disliked private power companies, which he branded as organized "for ruthless exploitation, uninterrupted and unrestrained by anything approaching effective Government intervention and control."[6] Morris Cooke agreed with Pinchot. Cooke was an engineer but also a social activist who believed that electricity was so powerful a tool that it could lift up civilization, and certainly the American people. Both men firmly believed that the Giant Power proposal was a plan "by which most of the drudgery of human life can be taken from [the] shoulders of men and women who toil, and replaced by the power of electricity."[7] Both men believed that a national grid was inevitable and that the question was who would control it—the public or the private utilities? Cooke was committed to bringing electricity to the farms of Pennsylvania, where only 12,452 of 202,250 farms had electrical service. About 20,000 generated their own power, but Morris discounted stand-alone systems, ignoring the *Agricultural Engineering* article suggesting that "the individual electric plant has been the means of introducing electric service into many thousands of farms in the United States by providing energy for lighting and small power purposes."[8]

Giant Power, proposed as a state agency, would charter existing companies and would also authorize new transmission companies to construct lines and then transport electricity. These transmission companies would sell power to distributing companies (already existing), except for new service. Companies were expected to make a profit, but they would all be under the umbrella of Giant Power. It was complicated (as electrical systems are), and it was vigorously opposed by private companies. It smacked of socialism and nationalization at a time when free enterprise was ascendant. The legislation failed to pass in the Pennsylvania General Assembly.

Although Giant Power failed, many of the ideas of Pinchot and Cooke became part of the electrical system called the grid. Furthermore, the idealism of the Giant Power plan, that "either we must control electric power, or its master and owners will control us," should not be forgotten.[9] Since the time of Abbot Samson in medieval England (chapter 1), power monopolies have impeded people's freedom, re-enslaving them through high rates and arbitrary decisions. Properly regulated, electricity could free men and women from the drudgery of hard labor and enrich their lives.

Pinchot and especially Cooke wanted every farm in America to have electrical power. However, private companies were interested in profit; thus while urban homeowners enjoyed electricity, those who lived on farms remained in the dark. Cooke got his chance when President Franklin Roosevelt put him in charge of the newly created Rural Electrification Administration in 1935. As we have seen, the growth of the REA proved a disaster for stand-alone wind energy units, but it was a popular program that did bring the grid to rural America.

In spite of what Pinchot and Cooke believed, their prediction of a national grid never materialized. National grids do exist in Great Britain and Germany, but the United States was simply too big, with state jurisdictions and entangling transmission systems so complex as to frustrate the finest engineers of the day. What evolved instead were the electric-utility holding companies. At one point, seven holding companies controlled 40 percent of electrical power in the nation. The great entrepreneur Samuel Insull headed the Middle West Utilities Company. From his Chicago headquarters he controlled 8 to 9 percent of the nation's power. By 1930 these huge holding companies came under increasing attack for their monopolistic practices. After the financial collapse of Insull's company and his conviction of mail fraud, it was time for a change. In many ways, the Depression of the 1930s cooled the ardor for holding companies and led the way to federal legislation.[10]

In 1935 Congress passed the Public Utility Holding Company Act (PUHCA), which attempted to bring some organization to a chaotic grid. The act and the Depression reversed the spread of utility holding companies and increased the involvement of federal government. In 1933 Congress had created the Tennessee Valley Authority, and by 1935 the agency was building its first dam. In the countryside the Rural Electrification Administration was forming farm cooperatives and bringing central power to hundreds of thousands of farmers. These actions broke the virtual monopoly of private utility companies and led to what we have today: a blend of private utilities and publicly owned and operated companies.

The First Commercial Wind Turbine Connects to the Grid

Although many thousands of small turbines spun throughout the West from 1920 to 1950, none were connected to the grid. The direct current (DC) wind turbines were not compatible with the grid, which operated on an alternating current system. But one very large wind turbine could work compatibly with the grid: Palmer Putnam's brainchild (chapter 1). Beyond its size, the genius of Palmer Putnam's pioneer venture was in devising a system to feed the turbine's DC current into the Central Vermont Public Service Company AC current. Engineer John B. Wilbur attempted to explain the accomplishment in lay terms: It "requires careful speed regulation, since it becomes necessary to drive the AC generator, which in this instance is synchronous, at a constant speed, while wind conditions are varying, in order that the frequency of the system to which the wind turbine is connected may be matched."[11] Putnam and his team collected a disparate, chaotic energy source and integrated it into a steady, orderly system—an engineering feat that required skill and patience. It was an absolutely necessary advance for the day when thousands of huge turbines would feed their product into the grid.

The PURPA Revolution

The Smith-Putnam turbine was an experiment. The Central Vermont Public Service Company, while participating in that experiment, had no interest in building its own turbines or accepting electricity from wind sources. Utility companies, whether private or public, operated as regulated monopolies. They saw no benefit in opening the grid to *any* alternative energy production, especially when generated by independent producers. In regard to their product, they operated in what Jason Makansi calls "the traditional way." In his words that meant "Burn It, Convert It, Move It, Distribute It, and Then Consume It."[12] All this was accomplished by one vertically integrated company that controlled the process every step of the way. Any effort to break this closed system was met with a barrage of obstacles designed to frustrate and discourage.

If utility companies had been open to developing renewable energy capacity for their closed systems, they could probably have maintained their stranglehold on production and distribution of electricity to their customers. However, at the walnut wood–paneled board rooms of PG&E and Southern California Edison there was only a cursory, lip-service acknowledgment of the nation's increasing dependence on foreign oil and lack of energy production diversity. Discussions of clean, renewable energy technologies never made it onto the agenda.

The structure began to break down in the 1970s in spite of utility company resistance. Deregulation of other monopolies, such as the airlines and communications, suggested that it was time for a change with utilities. Furthermore, for the first time the federal government funded a little research and development with renewable resources, such as the wind. When Jimmy Carter came to the presidency he gave added weight to renewable energy sources, as he declared the nation to be running out of oil and natural gas, creating both a crisis and a mandate. His appeal fell on deaf ears; certainly the utility companies were singularly uninterested. They wanted no legislation that would break their monopoly on electricity production and open the grid. In other words an entrepreneur foolish enough to build a commercial wind farm would have no market for the product. This changed in 1978 with congressional passage of the National Energy Act. One of its five parts was labeled the Public Utilities Regulatory Policy Act (PURPA), legislation that reflected utility company interests and was quite unintelligible to the public. It would have been unmemorable (and incomprehensible!) except for section 210, titled "Cogeneration and Small Power Production."[13] The act defined cogeneration as the use of relatively small turbine generators (gas or steam) that could utilize discarded heat, as might be found in a paper plant or a petroleum refinery. More germane to our subject, "small power production" referred to solar power (wind, sun), biofuels, and waste, geothermal, and hydro power. The guidelines called these units "qualifying facilities" (QF). The act required utility companies to provide service to these facilities and—more revolutionary—*purchase electric energy from these facilities*. Payment would be based on "avoided costs," the costs a utility would have incurred if necessary to buy the power from another source, or the cost of constructing and maintaining a power plant if the QF did not supply the power.[14]

This legislation drew little attention from the utility lobby. Perhaps they did not realize the implications, or if they did, perhaps they believed that any such regulations could be emasculated by their battery of high-priced lawyers. They misjudged. The PURPA law was the most important act ever enacted for the wind energy business. Quite simply, without the law there would not be commercial wind energy in California or anywhere else in the nation. Once utility company lawyers understood the significance of the law, they did their best to neutralize it through weak regulations. But in spite of repeated attacks, PURPA has survived, and the Federal Energy Regulatory Commission has administered the law in the ways Congress intended. In other words, the law truly opened the electrical grid to small producers and stimulated new forms of energy production. The 1980s saw many fights about PURPA. Perhaps symbolic was the fight of the residents of a New York City building on lower East Side. They raised a tiny Jacobs 2 kW wind turbine on the building's roof and then demanded the right to feed their wind power into the Consolidated Edison grid and be paid. The powerful utility refused, but the New York State Public Utility Commission sided with the residents. Con Ed would have to allow the hookup.[15] Backed up by federal law, David defeated Goliath.

Although less dramatic, more significant was the remarkable turn of events in California. Beginning in 1980, the combination of the PURPA law, state incentives, federal tax credits, plus accelerated depreciation on turbines made wind energy an attractive investment. It may not have guaranteed a profit, but the incentives insured against loss. What emerged in California from 1980 to 1987 were the first and the most ambitious wind farm projects the world had seen. I have described this growth in chapter 1; here I must underscore that for the first time significant wind energy was fed into the California grid system, purchased mainly by the Pacific Gas and Electric Company and Southern California Edison.

Thus by the 1980s the right of wind energy companies to market their product on the grid at a fair price (avoided cost) could not be disputed. Significant amounts of renewable energy became available to homes throughout the state. However, although the right of wind companies to "wheel" electricity on the grid went unquestioned, the value of that wind energy varies. For instance, on a typical California summer afternoon the peak demand may be at four o'clock, the time when air conditioners are at work. It is also a time when cool air from the coast flows inland to fill the vacuum caused by hot air rising in the Central Valley. Thus the value, or what experts call "capacity credit," is very high. At night the capacity credit might be divided by three. A grid system must be able to meet the peak demand, and wind energy varies in value according to the need. If the wind turbines harvest night wind, the electricity does not have great value. It is a problem. It would not be if we could *store* electricity to use when needed. Progress in storage technology is discussed later in this chapter.

Grid Limitations

Grid managers or gatekeepers worry about instability arising from the fact that wind energy is not constant. To use the human circulatory system as an analogy again, nurses can remove a pint of blood without any serious consequences. But inserting a pint of blood into the body of someone who does not need it might be a different story. Inserting electricity into an already full grid can cause precariousness. Engineers believe that more than 10 percent of wind energy will cause instability.

Gauging the circumstances is complex. Transmission lines come in different sizes, and so do their towers. Short lines may be 115,000 or 138,000 volts (more conveniently expressed as 115 or 138 kilovolt lines). More common for longer lines, at least on the West Coast, are 230 and 500 kilovolts. The largest high-voltage lines, primarily on the East Coast, are 345 kilovolts, with a few at 765 kilovolts. The current flows through aluminum transmission lines; although copper is a better transmitter of electrons, it is much heavier. Once the electrons leave the generator they enter a substation, where they may be converted to DC power, which is common today for long distance transmission. When the power reaches its destination, usually an urban substation, operators convert it back to AC power for our consumption.[16] It then goes out to distribution centers and into our homes.

The problem with wind energy is that it must enter the grid immediately, because under the conditions of PURPA the power grid must accept renewable energy first. It has priority. Yet renewables do not have the right to displace existing grid users. Thus base load generators, such as those delivering coal and nuclear power, generally have priority. What we have here is an enigma: a contradictory policy. Grid operators must do a constant dance of adjustment, giving balance to the grid. If the base load provider, usually a coal plant, is duplicating the production of the wind turbines, and the combined power output exceeds the capacity of the grid, the grid operator can require the base load provider to ramp down to 50 percent of its capacity or less. This takes time, and fluctuations in output waste energy. Normally the grid master has some leeway and can balance the load. If the system is strained, often the substation can take care of the overload. Some substations have "choke coils," which can deal with a power surge or malfunction by choking it off "like a constriction in a garden hose."[17] And somewhat like the circuit breakers in homes, an overload trips the breaker automatically to prevent damage to or fire in the substation.

As yet, this inefficiency has not been resolved. Further exasperating the situation is that in Texas at least 10 percent of the grid must be free for emergencies.[18] If power must be transferred from one region to another, transmission capacity is often unavailable. For instance, Ryan Segley, project manager for Crownbutte Wind Power in North Dakota, stated that the company is waiting to see if there is enough room on transmission lines to go ahead and build: "We're waiting for the queue process to see if they can get us on the transmission lines." Segley caught the reality of many wind development projects when he stated: "You basically work back from transmission capacity or what you think is the transmission capacity for that area."[19] If there is no transmission capacity, it is foolish to build.

The answer is to build more grid capacity. But that is expensive, and it is consumers who will foot the proposed $5.75 billion bill to bring the Texas grid greater capacity. Environmentalists oppose construction of new transmission lines on grounds not much different from their opposition to turbines. The Southern Utah Wilderness Alliance is fighting a transmission corridor along Utah's Moab Rim. In Nevada, there are proposals to build transmission lines across the Desert National Wildlife Refuge Complex. Bill Huggins, representing the Friends of Nevada Wilderness, responded at a Department of Energy public meeting, "absolutely not."[20] Generally, the public does not want new transmission lines criss-crossing the nation like a crazy quilt. Yet, conservationists like Huggins are in an uphill battle when fighting green power transmission. As journalist Judith Lewis notes, coming out "against a project that brings wind energy down from Wyoming [means] you come out against polar bears and in favor of cataclysmic drought, all to prevent a localized disturbance in your backyard. No matter how pristine that backyard, or how many rare species it contains, saving it can't possibly trump saving the coasts from rising seas."[21]

As Texas oil man Boone Pickens moves into the wind energy business, it is unlikely that significant opposition to new transmission lines will emerge. He will pay for the hundreds or thousands of turbines he plans to erect between West Texas and the Canadian border, but who should pay for the expensive upgrading of the grid system? Perhaps it makes sense to send the bill to the ratepayers or consumers. But returning to the idea of subsidies (see chapter 5), the United States has provided generous subsidies for what Congress and the people consider to be of national significance. There are many reasons for rebuilding the grid system. It is generally believed that the nation's system is antiquated, becoming dangerous, and certainly prone to failure. From the perspective of global warming and our increased concern about the environment in which we and our children must live, it is time to begin rebuilding. Some people favor spending the nation's money to put a person on Mars or to conduct foreign adventures, but from the point of view of the health and welfare of the people, it is time to fund rebuilding of the nation's grid with the aid of the federal government.

We should get to the job soon. One of the great advantages of wind energy is that units may be put on line within two years, or even one.[22] However, that advantage is lost with regard to the power grid. Thomas Friedman tells the story of efforts by Southern California Edison (SCE) to build a $2 billion power transmission line to bring wind power from Tehachapi Pass to Los Angeles, a distance of 275 miles. First the company had to clear the hurdles of the transmission planning process. It had to determine a need and work out who would cover the expense. All players had their say, from the Federal Energy Regulatory Commission to the California Independent System Operator and the wind developers. Once it was determined that SCE would pay for the line (eventually charging customers) and all voices had been heard, two years had passed. The next step was an extensive environmental study, dealing with vegetation disturbance and possible endangered species on both private and public land (U.S. Forest Service, Bureau of Land Management). The California Public Utilities Commission took more than eighteen months to review the proposal and do another environmental review, and so on. Ron Litzinger, the senior vice president for transmission and distribution for SCE, said: "We started the project in 2002, and today [February 25, 2008] we have permits for one-third of the project in hand." He estimates that the transmission line will be operational in 2013—after eleven years.[23]

No one wants to undercut environmental processes, but accelerating them can have benefits. Governor Arnold Schwarzenegger, in his efforts to move California toward a sustainable energy future, underscored this: "Environmental regulation is holding up environmental progress."[24] Procedures in California are the antithesis of those in Texas; middle ground might be wiser in both places.

A National Grid

For a model we may use the federal highway system, authorized under the Federal-Aid Highway Act of 1956. Certainly it was lobbied for by automobile manufacturers, but just as certain was a clear need as Americans rejected public transit and chose the automobile as their method of transportation. Between passage of the act and completion of the system the country spent $425 billion (in 2006 dollars); few criticize the investment. Fifty-six percent of the cost was borne through user fees such as the gasoline tax, and the remaining monies have come from the federal budget.[25]

Yesterday we were wed to the automobile. Today we are earnestly seeking divorce, and tomorrow we will re-wed to electricity. As the nation and the world decrease their carbon footprint we will increase our use of electricity. But first we must use electricity efficiently. Second, that electricity must be produced through such renewable methods as geothermal, solar, wind, and biomass power. We will require base load generation, and it will probably be nuclear. But whatever the method of production, we must have a way to get the energy from production to consumption. The grid is our method of distribution, and it needs a massive upgrade. Recently Robert F. Kennedy, Jr., called for a massive overhaul of the grid to include greater capacity, greater efficiency (use of DC power for long distance transmissions), and opening of the grid to more local and national green power producers. If we are ever to escape our carbon dependence, this massive change must be accomplished. It will not be easy and it will be costly. Kennedy estimates that "construction of efficient and open-transmission marketplaces and green-power-plant infrastructures would require about a trillion dollars over the next 15 years."[26] The U.S.

Department of Energy suggests that the part targeting wind energy would be about $20 billion, but that is probably a low estimate.[27] Part of this immense cost would be offset by state and local government and private entrepreneurs, but as with the federal highway program, taxpayers would shoulder much of the burden. Is a national effort to shift to renewables and rid the nation of dependence on oil and carbon-based fuels worth the cost? Think of the benefits in foreign policy, trade deficits, independence from oil enslavement, lessening of global warming, and finally, individual and national health. For many Americans, the cost-benefit ratio favors the benefits.

Wind energy presents unique problems for the grid. Jason Makansi, author and electricity consultant, notes that electricity from "renewable sources is unpredictable and inconsistent. For this reason it has a low value with respect to grid operations, even though it may have high value with respect to the environment." But the problem of wind energy and the grid could be resolved, Makansi believes, if we could *store* electricity and inject it into the grid when needed. Almost every commodity people manufacture increases in value and utility if we can store it until needed. Storage "makes renewable energy matter" by balancing "temporary dislocations in supply and demand."[28] Makansi believes that the storage goal should be 15 percent of the nation's generating capacity, which is about the level of storage in the natural gas industry.[29]

But immediately the question arises: how do you store large amounts of electricity? It is not easy. Organizations such as the Electricity Storage Association are addressing the problem, but storing 15 percent of the nation's generating capacity is highly unrealistic. Today the most common method of storing electricity is through hydropower. Pumped hydro uses two water reservoirs, separated vertically. At night, when demand is low, operators pump water to the upper reservoir, and in afternoon, usually at peak load, the water is released into penstocks and turbines. Operators make money by pumping when the cost per kilowatt-hour is low, say 5 cents, and releasing when the rate is high, say 10 cents per kWh. There are thirty-eight hydro storage units in the nation, a minuscule number.[30]

There has been a great deal of research into storage batteries, of course, and the most promising are the NAS batteries, which would store reasonable amount of electricity at night to shave the peak load during the afternoon. But perhaps most interesting is a project under way in Iowa.

The Iowa Stored Energy Park, now under construction, will feature a wind park of 75 to 150 MW. When demand is high and the wind is blowing, the electricity will go directly into the grid. However, when the wind is blowing but demand is low, the electricity from the wind will drive a large air compressor. The compressed air will be inserted into a sandstone cavern where it will remain under pressure. When electricity demand is high, compressed air will be released and mixed with natural gas to drive a conventional combined-cycle turbine. The engineers expect that the air–natural gas combination will be 38 to 64 percent more fuel-efficient than a turbine without compressed air. This compressed-air energy storage (CAES) system should be in operation by 2015. It is being funded by more than one hundred municipal utilities in Iowa, Minnesota, and the Dakotas, with additional funding from the federal Department of Energy. Depending on the final size, designers expect the facility to produce 200 to 300 MW at capacity. If this technology is successful and there are sufficient sites available, this could be a true breakthrough. As the designers state: "Now, by storing the wind's energy for when we need it, wind is something we can control and count upon."[31]

Changing Consumption Patterns

Certainly there are other ways to control the peak loads of our grid system. Utility companies have also assumed that consumers should be given all the electricity they want. And generally, the assumption has been that the more a customer consumes, the cheaper the per kilowatt price. This flies in the face of sensible environmental stewardship and embraces the nineteenth-century American myth of superabundance. We have been, and still are, a wasteful society, believing that we can consume as much electricity as we want, as long as we can pay for it. The wealthy tend to have an immense carbon footprint and to be more than willing to pay for it. Does it make some profligate consumption ethical? Probably not. In the United Stated we are not about to set up a moral and ethical department to police energy consumption. What we can do, however, is make it expensive to consume more than a reasonable number of kilowatt-hours. More and more utility companies have scrapped obsolete rate structures that reduced rates for high consumption. They now have a straight rate structure price, and a few utility companies have upturned capitalist principles by *increasing* the rate as a consumer increases use. Such a structure must be equitable but must also consider our environment as a key, and equal, player.

Just as important as the energy rate structure is the way utilities measure electricity consumption. Cities throughout the nation, and particularly in the arid West, have placed restrictions on water use, particularly during the summer months. We now need to do this for electricity. However, because of the nature of electricity and the fact that we use it all day, every day, we must approach the problem in a different way. The power company cannot turn off our meter without seriously disrupting our lives.

The answer is "smart" meters. These measure the amount of kilowatt-hours a household uses at particular times of a twenty-four-hour period and assign a price. For instance in the afternoon, say from 2:00 to 6:00 P.M., the meter would record use at 18 cents per kilowatt-hour. From 6:00 to 11 P.M. it might drop to 14 cents. At night, from 11:00 P.M. to 7:00 A.M., the rate would be reduced to 9 cents—one-half the peak period price. Essentially the smart meter would mimic the price of the wholesale market (often called the spot market), and consumers would be in charge of the price they pay for electricity. Armed with this information, I am convinced that consumers would alter their behavior, doing laundry and dishes at night and programming the meter to cool the house during off-peak hours. If we can cut down electricity use during peak periods, the spot price will drop. "Everybody will save," states Ashok Gupta of the Natural Resources Defense Council.[32]

We should not assume that these sophisticated meters are new and untried. Between 2000 and 2005 EnelSpA, the dominant utility in Italy with over 27 million customers, installed these meters across its entire customer base. They are working well and testify to the power of next-generation metering systems. In California the Pacific Gas and Electric Company, a leader in the wind energy field, is installing 9 million smart meters that record electricity consumption on an hourly basis. A number of other countries worldwide are experimenting with or installing such meters.[33] All of this bodes well for increased use of wind energy. We may not be able to alter the habits of consumers through environmental or ethical arguments, but pure economics represented by these meters will make a difference. If we really want to resolve the grid problem in a sensible way, and one that will encourage more use of wind energy, we will shave those energy peaks into hills and the chasms into gentle valleys.[34]

Experts and utility managers have often said the wind generates electricity at night, the very time we do not need it. But perhaps it is time for us to adopt to nature's requirements, rather than the other way around. Imagine a plug-in electrical car that gets its batteries charged at night. This technology is coming soon, perhaps sooner than this book is published. Again it will smooth out the peaks and clearly increase the value of wind energy.

CHAPTER 5

Can the Government Help?

Subsidy: a direct pecuniary aid furnished by a government to a
private industrial undertaking, a charity organization, or the like.
—*Random House Dictionary of the American Language*

It was a clear, beautiful July day on Cape Cod, Massachusetts—with a
gentle wind blowing—when I made my first visit. I had been invited to
participate in the National Public Radio show *Justice Talking*, featuring a
debate over the Cape Wind project on Nantucket Sound. Out on Horseshoe Shoal, some six miles off the cape, developers proposed 130 giant
turbines, the first major renewable energy project on the East Coast and
the first offshore wind installation in the United States. Before an audience of two hundred vocal citizens, John Passacantando, executive director of Greenpeace, and Jerry Taylor, a senior analyst for the Cato Institute,
squared off. I provided historical background.[1]

The debate was lively, and with a partisan crowd it was an entertaining
evening, whether you were for or against. Having introduced the Cape
Wind project in chapter 2, I return to it here to address the role of the government in fostering wind energy. Jerry Taylor was emphatic in his belief
that if all the local, state, and federal incentives were stripped away, there
would not be any commercial wind farms in the United States. Without
federal tax breaks and subsidies, wind energy development would not exist.

In other words, in a free market economy, wind energy could not compete. Whether Taylor is correct is problematic. The wind energy industry today receives impressive incentives, quite a change from before 1974, when no government support existed and no wind energy either. Certainly much of the success of the industry today can be attributed to the encouragement of the federal government.

A Suite of Subsidies and Incentives

Glenn Schleede, a dedicated opponent of wind energy, has been a vocal critic of just about all aspects of the industry. He is most effective in enumerating the large tax advantages and the various incentives the wind industry enjoys. They can be divided into six categories. First, we can examine the concept of *accelerated depreciation*. Every power production company enjoys depreciation, and so do many people who have a small business or even claim a home office as an income tax deduction. For most power plants the deduction is spread over twenty years, but that is not the case with wind energy companies wishing to shelter income. The IRS allows them to depreciate their capital costs over five years, and often this devaluation carries over to state income tax returns. In fairness, the wind industry did not invent accelerated depreciation.

Second, and particularly significant to developers, is the *federal production tax credit* (FPTC). This subsidy allows a federal tax credit to turbine owners of approximately 2 cents per kilowatt-hour produced. This does not sound terribly significant, but it amounts to many millions of dollars per year for large producers. Schleede states the FPL Group (a subsidiary of Florida Power and Light and one of the largest wind plant owners in the nation) paid no income tax in 2002 and 2003 while reporting a net income of more than $2 billion. Over the years the FPTC subsidy has been controversial, but without it there would be little new wind energy development in the United States. In the past twenty years the production tax credit has acted as a barometer of the industry's health. When Congress has cut it off, making it unavailable for new turbines, construction has dropped off to practically nothing. When the credit has been available, turbine construction has surged.[2] With the 2008–2009 downturn in the economy, Congress passed the American Recovery and Reinvestment Act in early 2009. Under the act legislators renewed the FPTC for three years (2009–12), the longest period in the history of the subsidy. Also, companies that can no longer profitably use the production tax credit are able to take an energy tax credit for wind facilities placed in service between 2009 and 2012, but only in lieu of the FPTC.[3]

When President Obama signed the stimulus act on February 17, 2009, the wind industry hailed it as a positive signal; just how positive remains to be seen as administrators write and interpret the regulations. However, it does appear that the administration will reserve $1.6 billion for *renewable energy bonds*, a kind of alternative to the FPTC. These bonds could be crucial for Indian tribal governments and independent wind cooperatives (nonprofit), which because of their unique tax situation cannot profit from the federal production tax credit. The bonds may make it possible for them to raise the capital needed to get their proposed projects built.[4]

Accelerated depreciation and tax credits are the two primary federal incentives. A third important boost comes at the state level. Twenty-five states have adopted *renewable portfolio standards* (RPS). The typical portfolio standard establishes a minimum percentage of energy that must come from renewables such as wind, solar or hydro power, or biofuels. As such, adoption of RPS encourages wind energy since there are few hydro electric sites available and solar is, as yet, costly. RPS requirements are not a direct subsidy but rather a state policy that encourages the wind industry. The industry is lobbying for a *national renewable portfolio standard*, and at this writing the Senate Energy and Natural Resources Committee has drafted legislation. Under the bill the national goals for electricity from renewable sources would gradually increase to 4 percent in 2011–12, 8 percent in 2013–15, 12 percent in 2016–18, and 16 percent in 2019–20, reaching 20 percent between the years 2021 and 2039. Senator Jeff Bingaman, the energy committee chair, is eager to push the legislation. Like grid improvement, passage of this legislation may prove possible as a result of the fortuitous combination of federal need for job creation and leanings toward a greener economy. As Senator Bingaman states, the standard would "spur the development of the national green energy economy, creating hundreds of thousands of jobs, many in rural areas."[5]

Fourth, another form of indirect encouragement, perhaps more properly called a state mandatory incentive, is *renewable energy credits*. When a utility company fails to produce its assigned portfolio requirement of renewable energy, the state requires that it buy renewable energy credits from renewable energy producers who have a surplus. These credits are sold by state auction, much as in an open market. The cost in Texas, for instance, has varied, but it is usually around 2 cents per kWh. Added to the federal production tax credit, a wind energy company receives about 4 cents per kWh. Is this an unfair advantage? Certainly not, for the more traditional sources of energy production have their own basket of subsidies and incentives from which to draw.

Fifth, many states authorize *voluntary green electricity programs* in which a consumer purchases a block of wind power at a premium price. Such programs are common in regions where there is only one utility provider. They are somewhat symbolic, but the added income for a renewable energy provider is real. In a few deregulated states, such as Texas, you can sign up for a total renewable energy program. Green Mountain Energy provides electricity generated solely from wind and hydro power sources at a price of 1 to 2 cents per kWh above the cost of power from more traditional sources, mainly coal.

At the county or municipal level we find a sixth type of incentives. Like any industry a wind energy company will try for a *tax abatement*. Some state governments have lowered or eliminated state and local property taxes. More common, wind farms developers negotiate voluntary payments in lieu of taxes. These payments are often for a specific municipal project to enhance the community, such as a community aquatic center. Wind developers generally prefer this option, for contrary to property tax, it is for a limited duration.

Are these various incentives defensible? Have wind developers inherited the mantle of the industrial exploiters of yesteryear? Have wind energy promoters misled the American public and its legislators? Has the industry overstated the benefits and understated the costs? It depends, of course, on who answers the questions. Today the wind energy industry rides the crest of a wave of popularity, a darling of the public as well as legislators. The industry has capitalized on a well-known fact: wind energy is the most environmentally benign method we have to produce energy. In a nation desperately trying to negotiate both an environmental ethic and a consumptive lifestyle, we treasure an energy source of moderate impact. Until recently the nation did not appreciate wind energy attributes. Now we do. Turbines may not be as productive as advertised, and yet it is difficult to discount an industry that offers us a brighter future.

Subsidies in Perspective

In early April of 2008 Congress issued subpoenas for five top oil company executives to justify their combined $123 billion profit for 2007. They asked the CEOs why Congress should continue granting them $18 billion in annual subsidies. The executives' explanations were lame, stressing the long-term nature of the industry and the huge investment requirements. As to renewables, Stephen Simon of ExxonMobil stated that his company had "studied all forms [of alternative energy], and the current technology just doesn't have an impact" that would lessen oil dependency.[6] The other oil company executives were more optimistic.

It seems incongruous that the five oil giants with incredible yearly profits would insist that their $18 billion subsidy should not be eliminated or suspended. Yet insist they did, and the oil subsidies will probably remain. Once the government extends a subsidy to an essential industry with a powerful lobby interest, it is difficult to take that subsidy away.

The attitude of the oil giants reminds us that a subsidy means different things to different groups. To the industry receiving the subsidy, whether it is direct, indirect, or for infrastructure, it will prevent the decline of that industry (think American automobiles). To the opponents of wind energy, subsidies constitute an unfair governmental action and are the antithesis of a free market economy. However, two facts must be considered. Some of those who cry foul about government subsidies for wind energy have been the recipients of subsidies for decades, as we see with the oil executives. They have a selective memory of history. Second, in weighing value, we are at last giving the environment fuller consideration in the economy. The environment should always have been part of any cost formula, a significant externality if you will, but the various price tags of pollution have been neglected until the last few decades. The complainers are analogous to the individual who bemoans high taxes for welfare programs yet drinks at the federal trough whenever the opportunity arises. Subsidies are all about paradoxes and, in many cases, denial of the realities of capitalism, politics, and power.

The example of big oil reminds us that tax subsidies or tax breaks are an integral part of the American democratic system and the capitalistic system. For individuals, tax breaks are something to enjoy; for companies they are all about Washington power and influence. There is hardly a major industry that is not constantly seeking an economic advantage from government. Wind energy has finally squeezed in, and critics object.

In a fairer and more perfect system, government subsidies would not exist. In 2002 the Sierra Club's Carl Pope and Ed Crane of the libertarian Cato Institute, two leaders with John Muir and Adam Smith as their respective heroes, joined hands to call for the end of *all* subsidies in the budget and the federal tax code.[7] Along the same lines, when it came to subsidies at the Nantucket debate between Jerry Taylor of the Cato Institute and John Passacantando of Greenpeace, the two agreed. Taylor vigorously attacked wind energy for its capacious subsidies. Passacantando agreed that subsidies should end, but with the proviso that the government's entire subsidy program must be terminated. The audience smiled, knowing that the end of *all* energy subsidies was not something that would happen in their lifetimes.

Margaret Adler, the National Public Radio moderator, interjected with figures showing that all of our energy sources receive huge federal subsidies each year: nuclear power, $4.6 billion; oil, $2.8 billion; and "clean coal," $1.6 billion.[8] In my own work I found that between June 1955 and June 1964, the Atomic Energy Commission and other federal agencies had invested $27 billion of taxpayer money in the development of the "atoms for peace" program, with only meager results.[9] Recently, the Renewable Energy Policy Project issued a report on federal subsidies. It found that of the $150 billion spent in subsidies for wind, solar, and nuclear power, 96.3 percent had gone to nuclear power. Put another way, since 1947 cumulative subsidies to nuclear power have cost each American household $1,411, compared to $11 for wind.[10] Carl Levesque, a wind energy analyst, noted that the 2008 U.S. budget called for research and development funding for nuclear at $547 million, while wind energy research and development was at a paltry $38.3 million.[11] Of course this figure did not take into account the substantial production tax credit that wind producers received.

To put wind energy subsidies into historic perspective, it is helpful to look at the role federal subsidies have played in the development of the nation. If the past provides us any lessons, there is a place for subsidies. In the early nineteenth century states helped fund canal projects on the eastern seaboard. During the 1830s President Andrew Jackson was a strong proponent of subsidies, which in that decade were called "internal improvements." Better known were the massive federal subsidies to build railroads, particularly the transcontinental railroad constructed by Irish and Chinese labor in the 1860s. The Union Pacific and the Central Pacific received generous grants of land and loans simply because the federal government considered it vital to the development of the nation to have an improved transportation system tying the East Coast with the West Coast. Historians still debate the efficacy of the railroad subsidies, but at the time workers would have laid very little track without them.

Federal subsidies have always been available when Congress and the president consider projects to be in the interest of the nation. Recently massive subsidies in the form of loans have been injected into the nation's financial markets and the automobile industry, in both cases to prevent catastrophic decline. Perhaps more germane has been the interstate highway system. Pride of the Eisenhower years, the system represents a model of massive federal assistance to establish something we accept as essential. Today we have a fine national highway network, but production and transportation of electricity have fallen behind. If the country wants a modern, dependable, and environmentally sustainable electrical system, federal infrastructure support should not only continue but increase. A sympathetic public policy is the engine that makes wind energy function and grow.

Federal Research and Development

I would be remiss not to mention the assistance of the Department of Energy's Wind Energy Program, and particularly the National Renewable Energy Laboratory (NREL) at Golden, Colorado. The NREL library, open to the public, makes no pretense of holding historic materials but is useful for contemporary documents. The mission of the Wind Energy Program is to increase the viability of wind energy, particularly through public-private partnerships. In the 1980s the Department of Energy and NASA were not successful in development of a large commercial wind turbine, but the research programs brought new knowledge and reduced the cost per kilowatt-hour of wind power. Today the National Wind Technology Center (NWTC) does a great deal of testing, often of ideas and equipment introduced by wind turbine manufacturers. It represents a technology enticement for this relatively new industry. The center assists the private sector in creating advanced wind systems, whether through integration into the grid, blade research, or small turbine development. It is a clearinghouse for new ideas on a host of topics.

Does the National Wind Technology Center represent a federal subsidy to wind plant manufacturers and developers? It certainly does not discriminate against private companies. It is willing to investigate scientific and engineering problems that private enterprise would be reluctant to take on. The administrators of the program make no effort to apologize for the public-private connection. According to their own literature the NWTC "is designed not only to be a center for research, but a technology magnet for a new industry. It is a place where NREL researchers work side by side with wind turbine developers to create the advanced wind systems of the future."[12] Yet government workers represent the interests of the nation and the national energy issues are center stage. Perhaps the support given the wind industry could be considered an indirect subsidy. Times are changing, and the kind of support that the federal government offers the industry would have Palmer Putnam and his friends spinning in their graves.

State Aid

Regulation of the electrical energy industry has traditionally been a function of state government. From the Progressive Era onward, state regulation has, in theory, protected the consumer from the excesses of monopoly. Evidence of state support or neglect is obvious in the wind development of the different states. California was the first state to make an effort to attract renewable energy, particularly wind power. Governor Jerry Brown (1975–83) was determined to pursue an energy policy emphasizing renewable sources of power. To do so he created the Office of Appropriate Technology and asked Sim Van der Ryn and Ty Cashman, both environmentalists focused on sustainability, to head the agency. These leaders successfully lobbied for state tax credits and encouraged the wind industry in multiple ways. Also on board the California effort was the Public Utility Commission (PUC), committed to funding various studies to assist and encourage wind energy companies. The PUC often cleared the bureaucratic path with what we might now call "fast tracking." With such a proactive policy the state of California created over 90 percent of the nation's wind energy between 1980 and 2000.

Chapter 2 recounts the Texas effort to enter the business and the remarkable acceleration that presently puts it well ahead in installed capacity (7,116 MW). But the most remarkable growth in recent years has been in Iowa. A state best known for corn production has found that wind energy and agriculture can coexist harmoniously. Iowans found an opportunity for multitasking on their vast farm lands: that is, farmers can profit from what comes from the land as well as from the steady breeze that caresses the rolling landscape. The result has been exceptional. The most recent AWEA state statistics list Iowa with a wind capacity of 2,790 MW, surpassing California's 2,517 MW.[13]

Encouraging this development has been the Iowa Energy Center, located in Ames near the state university. The center receives funding from an "annual assessment on the gross intrastate revenues of all gas and electric utilities in Iowa." It provides low-interest loans, supports research and demonstration projects, encourages net metering (whereby the meter runs forward when using power from the grid but backward when the turbine sends electricity into the grid), and above all has created accurate state wind maps. These maps often show the way to potential sites. This kind of cooperation has resulted in projects such as the Pioneer Prairie Wind Farm near LeRoy in northern Iowa. In September 2008, some five hundred farmers and officials gathered in the middle of a cornfield to dedicate the first phase of the 300 MW project. The Bandaras, a singing group, played traditional western music, and according to the local paper, the event was a real celebration with everyone attending in high spirits and enthusiastic about the new turbines.[14]

Other states have found that state subsidies are not only good for the environment—they can also spur business. The state of Oregon, for example, realized the value of wind energy quite late but has made up for lost time. Its energy tax program began as a small, targeted program aimed at conservation and renewables, and in 2007 the legislature expanded it. The new rules allow companies to apply for up to 50 percent of the cost of a project (up to $20 million) in tax credits, if the company can prove that the project will save electricity and/or produce renewable energy. The handouts come from Oregon's Business Energy Tax Credit program, which has become the state's fastest growing tax shelter.

Some critics have questioned whether state tax credits are needed for successful wind energy projects, especially since companies that qualify are allowed to sell the tax credits. Such a program can be used to profit from projects that have little to do with the growth of renewable energy. For instance, detractors point out that the Klondike Wind Farms, centered in a windy location in eastern Oregon, has little need for subsidies yet seeks $44 million in state tax breaks.[15] In the eyes of critics, such an outlay of public money, especially in a struggling state, is unjustified. The Oregon legislature is expected to revisit the program.

State Renewable Portfolio Standards

We need to explore state renewable portfolio standards further, for they are among the most successful blueprints to encourage renewable energy development. Often traditional utilities balk at developing renewable capabilities. They prefer to stay with the tried and true, namely baseline coal plants. However, with the advent of RPS, they have no choice but to develop a percentage of their electrical power from renewable sources. If they do not, they will be fined or forced to purchase renewable power on the open market. As noted, twenty-five states now have portfolio standards. For instance, New York has a goal of 25 percent renewables by the year 2013, while Montana's objective is 15 percent by 2015. Tiny Rhode Island expects to generate 16 percent renewables by 2019.[16]

Will state renewable portfolio standards guarantee that the goals will be met? The answer is no, because most of the standards are goals, not mandates. Whether goals are met depends on the legislative commitment and that of the utility companies as well. Conversely, the absence of RPS does not necessarily signal a state's lack of interest in wind energy growth. The state of Wyoming has no RPS, and yet there is significant activity. At last count, the cowboy state had 676 MW of wind capacity and growing. Some states, such as Arizona and Nevada, have significant portfolio standards, but as of this writing neither has any wind capacity actually in the ground and producing.

As one state legislature after another adopts or increases renewable portfolio standards, their actions represent a commitment to renewable goals. They signal to wind turbine companies that the state looks favorably not only on turbines but on manufacturing facilities. New plants to produce turbines, blades, and towers have opened in Colorado, Illinois, Iowa, South Dakota, Texas, Arkansas, New York, and Wisconsin. Such factories springing up around the country have created some six thousand new high-paying green jobs, illustrating how new jobs are arising from the growth of the renewable energy business.

If state governors and legislators continue to boost the renewable portfolio standards, progress will continue unabated for the foreseeable future. It has almost become a state competition. On November, 17, 2008, California's governor Arnold Schwarzenegger threw down the gauntlet when he signed an executive order to increase the state renewable portfolio standard to 33 percent by 2020. For the governor of the most populous state in the union to announce a goal of *one-third* of electrical energy to be generated by renewable sources was truly remarkable. California already had an aggressive goal of 20 percent by 2010, but the new goal confirmed the leadership of the West Coast state and perhaps challenges Texas as a leader of environmental innovation. As a caveat, executive orders do not have the force of law over utilities. Yet government agencies like the California Public Utilities Commission must take positive actions to implement the new targets.[17] The commission has never been shy in accepting such environmental challenges.

With the downturn in the economy and many states desperate for new sources of income, we may see a state tax on the actual product of the wind turbines. A tax on electricity derived from the wind sounds somewhat bizarre. However, most states regularly tax the products of oil, natural gas, and coal. Electricity created by wind turbines may be next in line.

Local Support and Local Taxes

National and state support mechanisms are reasonably easy to define. Assistance on the municipal and county level is more variable. Every wind farm proposal must go through a local (county, township, town) permitting process. Depending on the state, county, and location, this can be difficult or easy. Some counties, eager for development and jobs, offer significant property tax breaks. Generally school taxes must be paid by a developer but county taxes are negotiable. Whether a significant subsidy is possible depends upon the mood of local residents and whether they view wind energy development as positive. The situation is much like that of any company coming into a county expecting concessions in exchange for creating jobs. Rural counties want development, especially jobs. Many rural areas, particularly in the windy Great Plains, have experienced population out-migration that they desperately want to stop.

However, because of the turbines' size and how they industrialize the landscape, wind installations usually produce controversy as well as tax breaks. In some cases, landowners who must live with adjacent turbines claim that the value of their land has diminished with the presence of the intruders. If a real estate appraiser agrees, the property tax must be lowered. Thus the property tax paid by a wind developer may be offset by the depreciation of adjacent land. Sometimes the only real winners are the attorneys who battle such issues in court.

Obtaining a building permit for a wind project becomes more tangled with county commissioners, usually three, often with distinct personalities and differing political convictions. By the time a wind developer approaches the county commissioners, the developer has usually signed royalty leases with local landowners and sometimes with absentee owners. Some wind developers, such as Babcock and Brown, are secretive about the conditions of their royalty and lease agreements and include confidentiality clauses that forbid landowners, under threat of a fine, from revealing the conditions of their leases. These clauses are solely in the interest of the wind power industry and should have no place in a negotiated contract. The North Dakota state legislature is considering banning confidentiality clauses because they deprive landowners of good information about the market price.[18] Imagine trying to price your home when surrounding sellers are banned under threat of penalties from revealing their own selling price. Each would have to guess the value of the resource, which is exactly what some landowners must do. Rumors ramp up, and feelings are ruffled. The sad facts are that some property owners are making plenty of money from the turbines, while others get nothing. It is a situation in which neighbors take sides, and they take their opinions to the county commissioners, and things can grow volatile. Fortunately, the robber baron days of wind energy developers are about over. Farmers and ranchers considering wind development now have access to information (usually free) to help with decisions. Furthermore, there are certainly ample attorneys who can aggressively represent a landowner's rights.[19]

Most county governments welcome wind development. The tax potential was first recognized in Palm Springs in 1989. At that time, Mayor Sonny Bono, the high-profile Hollywood star, vigorously opposed construction of seventy-four 500 kW turbines atop Whitewater Hill in San Gorgonio Pass. Bono announced that he would "fly to the nation's capitol . . . to do battle as Don Quixote did against windmills."[20] Secretary of the Interior Manuel Lujan assured Bono he would look into the situation. Lujan did nothing, which was just as well. Bono reversed his position by August 1990. The city was strapped for money, and Bono faced the disagreeable option of raising taxes. The companies, on the other hand, were paying taxes on $550 million of assessed valuation—but not to the city. From a public administrator's point of view, the turbines were a no-brainer. They did not require added schools, sewers, water service, police or fire protection, or paved roads. Admitting that he had been "pretty boisterous" against wind energy, Bono now recommended annexation of the wind turbines to add a largesse of $1.6 million in tax revenues. Soon the cities of Palm Springs and Desert Hot Springs were fighting over who could annex the wind farms.

Bono's dilemma has been played out hundreds of times in towns throughout the country. County commissioners, community leaders, neighbors, and occasionally even families must weigh the benefits of wind energy. Should the community allow these very visible turbines on to the landscape? Such a question can turn neighbor against neighbor and destroy a sense of community that has taken decades to mature. I have an acquaintance in Wyoming who bought land in Montana with the intent of erecting a dozen turbines. When he approached the county supervisors for the necessary permit, there was a public hearing in the small county seat. He was prepared to explain the project and its many benefits for the local community, and particularly the environmental and global good that the project would support. But he was totally unprepared for what happened. He was blindsided by a hostile audience, chastised, and called a "carpetbagger" and plenty of worse epithets. The local response shocked him. He found that global and environmental benefits do not mean much in a small isolated town. After the locals had their rant, the county supervisors ignored the opposition and voted three to zero to give him a permit. However, he felt so unwanted that he shelved the project for a year, or perhaps forever.

The uncertainties of local response have much to do with the geography and economy of the region. At Fort Stockton, Texas, town Director of Development Doug May welcomes wind developers and smoothes the way. Pecos County is desert and does not have many options beyond resource extraction. Oil, gas, and wind development are welcomed. They bring jobs and community amenities otherwise unavailable. Pecos County will do what is necessary to get turbines ringing the mesas, for their spinning blades mean a better life for the residents.

Again, there are trade-offs for rural counties. For instance, across the country in New York, the Maple Ridge Wind Farm yields $8 million annually in local tax revenue, 163 new jobs, and $1.65 million in lease payments to local farmers.[21] Do such benefits make the project popular? Not necessarily. As is explored in chapters 6 and 7, some community members are convinced that the benefits are outweighed not only by the visual and audible impact, but by the social friction that becomes unavoidable in small communities.

Voluntary Market Subsidy

Most wind energy companies also receive an indirect benefit quite independent of our normal definition of subsidies. Since wind energy has become environmentally correct, sympathetic private companies and residential customers are willing to pay a premium for electricity generated from the wind. Such companies as Intel, Pepsi, Dell, Whole Foods, and many others contract to purchase part or all of their electricity from renewable sources, mainly the wind. These days purchasing green power is a worthwhile marketing scheme. The companies can show their concern about global warming, and by their purchases they stimulate new wind energy projects. This kind of subsidy is in the *volunteer market*, where both companies and individual customers receive indirect benefits. The companies get positive publicity as responsible, progressive organizations working toward sustainability, while individual customers benefit from knowing they are doing something for the earth.[22]

But if we pay a higher price for a block of green energy, how do we know we are getting a return on our investment? In essence we must trust the utility. Sometimes that trust is not earned. In Colorado some forty-seven thousand customers signed up under the Windsource program to pay a premium to support alternative energy. After a state investigation, however, the Colorado Public Utility Commission found that Xcel Energy, the investor-owned utility, was selling more power under the Windsource program than it generated. Thus, Windsource customers were paying a premium price for nothing. Having been exposed, Xcel Energy will use shareholders' money to pay a proposed settlement of $2.6 million.[23] The customers will receive a refund, but one wonders just how many of these volunteer programs promise a product they cannot produce.

Feed Laws

There is another way for the national or state government to encourage renewables. If Congress passed "feed law" legislation, such an action would change the playing field for renewables. What is a "feed law?" We have to look to Europe, where wind energy leaders in Denmark, France, Germany, and Spain have fought for feed laws. What these laws do is guarantee the price a utility will pay for renewable electricity. Each law is unique to the particular country, but the producing wind energy owners will receive 80 to 90 percent of the annual average retail price. Thus a utility that charges its customers 10 cents per kilowatt-hour pays the wind energy developer 8 to 9 cents per kWh for the energy it contributes to the grid. These prices are somewhat complex and vary from country to country and region to region, but the important point is that renewable energy producers are *guaranteed* a fair price for renewables connected to the grid. Paul Gipe supports passage of a feed law in the United States, claiming that there would be no "cumbersome bureaucracy. No secret bidding. No sweetheart deals. One level playing field for all: farmers, homeowners, small businesses, municipal governments—everyone."[24]

If the nation is committed to increasing the percentage of renewable energy, it would be sensible to enact a feed law, particularly because it has been so successful in other countries. In 1990 Germany produced a minuscule amount of megawatts from wind energy. After legislators enacted the feed law, that figure rocketed to 12,000 MW by 2002. Paul Gipe does not mince words in his praise: "No other program has delivered more renewable energy than electricity feed laws. None. Neither net metering, nor renewable portfolio standards, nor tax credits, nor even PURPA has produced more wind-generated electricity than the feed laws used in Europe. To put it simply, feed laws work!"[25] As our nation struggles for energy independence and cooperation regarding global warming, perhaps we should examine the example of Europe more closely. Although we are reluctant to acknowledge it, the Old World has been outstripping the New in innovative laws and methods of dealing with the global environmental crisis.

Community Wind

There is an egalitarian argument for wind energy. After all, the energy source is free. Today, however, the industry is dominated by big business and is likely to be so for the foreseeable future. Its trade organization, the American Wind Energy Association, is largely funded by big business, while the turbine manufacturers and wind farm developers seek capital from such Wall Street financial firms as Bear Stearns and J. P. Morgan. Such a change represents the growth and maturing of the industry, and yet one feels a tug of nostalgia for the old leaders and companies that, while floundering, seemed to care more for bettering the environment than for profits. Most of the large turbine manufacturers and wind farm developers make claims about saving the world in their advertisements and lobbying efforts, but their objectives are all about the bottom line. The days of innocence are over.

Yet there is a movement called community wind. In northern Minnesota and Wisconsin independent cooperatives exist that offer a contrast to the huge Wall Street–funded projects. An organization called Windustry, headed by Lisa Daniels, and the Minnesota Sustainable Energy for

Economic Development (SEED) both have the kind of idealism and independence that provides a basis for renewable energy with a decentralized twist. They are reminiscent of the Grange movement a hundred years ago, when farmers fought the excesses of railroad monopolies. Community wind organizations favor local control and are determined that not all the monies generated from wind power will fly off to financial centers and mammoth utility companies. These two organizations do not oppose corporate wind farms, but they emphasize communal ways to promote energy diversity and efficiency. Windustry sponsors a network of farmers and rural people who search for reliable wind energy information. It seeks to empower communities to develop wind energy and work toward sustainability. But it is more than information. The organization encourages members of the local community to have a significant financial stake in any project. This is certainly possible on a small scale, although large wind farms, often requiring well over $100 million investment, are beyond the resources of a small community. In a sense, the wind industry imitates agribusiness, whereas community wind promotes the family farm. And as on family farms, money does not come easy. As Lisa Daniels mentions, because community wind involves nonprofit organizations, it is difficult for "local investors and community-based entities to take advantage of the main federal incentive for wind, the Production Tax Credit."[26] But with the favorable atmosphere toward wind energy and the community advantages (as opposed to resentment of utility companies and global developers), the new 2009 law can empower Windustry through a program of federal loan guarantees. It may be possible to fund large proposals. Public policy is the key, and Windustry does have the advantage of having the interest of the community at heart, rather than the stock dividend of a multinational company.

One hopes Windustry can expand its financial reach. Today, wind farm owners and local communities are often at odds. Everyone acknowledges the environmental importance of wind farms, and much of the discord surrounding the industry could be minimized if we could confer more, share local control as well as profit or loss, and view wind development as an investment in community.

Something of this shared community investment can be seen in Europe. The development of European wind farms from Holland to Denmark represents cooperation rather than competition. To understand Windustry we must look to Denmark, so long a leader in the movement.

The Danes have much experience with not only wind energy but also with how to promote diversity and community. For many years small farming communities formed neighborhood cooperatives, then purchased a wind turbine and distributed power for their own use, sharing in the profit or loss. Wind energy expert Paul Gipe has written that the "Danish wind turbine cooperatives and an association of wind turbine owners have had a profound effect on the development of wind energy in the country." He found that by the 1990s some 100,000 households (about 5 percent of the population) had an equity interest in a windmill guild or cooperative.[27] Furthermore, the Danes have incorporated energy efficiency and environmental awareness into their culture. They cajole their neighbors to stop wasting and go green. As one Dane put it, "To us, going for lower energy use is like a [competitive] sport."[28]

Denmark is not the only nation to allow small wind operators a chance. In Germany, according to engineer Martin Hoppe-Kilpper, farmers were responsible for at least half of the 250 MW surge in construction of wind generators between 1992 and 1997. They own shares in small wind companies. They are risk takers, and since they make their living from the land they are not adverse to landscape change. And they have been the recipients of subsidy programs for wind energy development.[29] If our government supported the rural community with half the effort of Denmark and Germany, such rural community efforts as are represented by Windustry may have a chance to provide a cooperative model. The desire and the organization are in place for a nonprofit body with a mission statement of environmental gain, rather than financial profit, to function in the market. If the interests of landowners and wind developers could be more in harmony, everyone would profit. What is needed is financial backing through a federal loan guarantee program.

The Arc of Energy History

For one hundred years the federal and state governments ignored wind energy, refusing to open their purses for any research or development. Wind energy was simply left to sink or swim in the world of free enterprise. Yet of course it was not free enterprise in the sense of a level playing field. Subsidized fossil fuels, nuclear power, and the federally supported REA hi-line projects forced people interested in wind energy to look for another line of work. It took a profound energy emergency to change accepted ways, and that crisis came on October 19, 1973, when the Organization of Petroleum Exporting Countries (OPEC) announced on an oil embargo to the United States. Within two months oil prices had quadrupled. Often gasoline was unavailable at any price as long lines of cars snaked around any gas station that had a supply. Six months later the embargo ended, but energy conservation had become part of the American lexicon. Millions of people became aware of the nation's vulnerability to foreign oil producers.

In this situation scientists and engineers returned to some old concepts, including harnessing the wind. William E. Heronemus, a nuclear scientist, led the way. He asked his fellow engineers at a meeting of the American Society of Mechanical Engineers if it would be an "unacceptable stigma to our society if we were to opt for an energy system whose science and technology would be very unsophisticated?" The answer was yes, but times were changing. A foreign-imposed crisis exposed the vulnerability of an energy economy based on petroleum. The predicament might have offered an opportunity for discussion of a national energy policy, but the Nixon administration fell back on the nuclear energy option. This promised power source, however, faced severe opposition on both an economic and ecological level. There was little choice but to consider renewable sources. The 1974 federal budget included $12 million for wind and solar research: a paltry sum, but these few crumbs of the pie ended the government's long-standing starvation policy.[30]

To have a real effect on the United States electrical energy mixture, Congress still had to break the old monopoly structure in which utility companies owned the means of both production and distribution. When Congress passed the Public Utilities Regulatory Policies Act (PURPA) in 1978, it represented a revolution. As we have see, the true significance of the act lay in the legal right of independent electric power producers to have access to the grid.[31] Until passage of PURPA, access to transmission lines by an individual or independent company was discretionary on the part of the utility. Utility companies wanted no part of buying electricity from individual producers and they did not accept PURPA gracefully. Those who attempted to tie into the grid faced policies designed to block entry. Yet the utilities could not skirt the law. The act withstood one legal challenge after another until it stopped being disputed.[32]

To return to the question of whether the government can help: it is a rhetorical question. Clearly federal and state support policies have made wind energy what it is today. Without such legislation there would be little renewable energy capacity in the nation. For wind energy to squeeze into the energy mixture took both direct and indirect government stimulation, and that encouragement changed the nature of electrical generation and gave birth to the wind industry.

CHAPTER 6

"Not in My Backyard"

Exoskeletal outer-space creations. . . . The once-friendly pastoral
scenes now bristle with iron forests.
　　　—Sylvia White, regional planner

To see [the turbines] as something beautiful. The breeze made
beautiful. The future made possible.
　　　—Bill Mckibbon

No sooner had the wind turbines gone up than environmental groups
registered strong objections. It was 1981 and the gentle, rolling hills of
Altamont Pass were changing dramatically. Wind turbines were invading by the hundreds. One of the first and most memorable descriptions
came from Sylvia White, a professor of regional planning, when she ac-
cused wind energy companies of "industrializing" the bucolic Altamont
hills with "exoskeletal outer-space creations." Employing anthropomor-
phic comparisons, she described the "long, sweeping blades attached to
what ought to be their noses . . . [with] legs . . . frozen in concrete, stationary but seemingly kinetic." The result of this alien onrush was that the
"once-friendly pastoral scenes now bristle with iron forests."[1] At the same
time, Mark Evanoff, director of People for Open Space, saw the turbines

as onerous as nuclear power, proclaiming that "we eventually will have to decommission windmills."[2]

Others saw the proliferating onslaught differently. Jeff Greenwald, a freelance writer, proclaimed that "they are not ugly. . . . One can almost mistake them for kinetic sculpture. Cows and horses graze among them; hawks wheel overhead." For him the turbines were "a far cry from what goes on around the business end of an oil refinery or a nuclear power plant." A comparative view like Greenwald's seemed to soften the impact of the turbines. Since those days of near thirty years ago, opinions on the wind turbines have not changed dramatically. Wind farms are still controversial and are the subject of heated debate across the country. I subscribe to a website that contains daily some fifteen to thirty articles of worldwide reach, mainly opposed to the expansion of wind farms.[3]

In short, opinions on wind energy are varied and often heated. Few people feel neutral. My wife and I once invited a well-known Western environmental writer and her father to dinner. When the conversation turned to wind energy, the father, a successful construction contractor, announced, "I hate them." His daughter, with a light touch, responded, "Dad, I love them." We steered the conversation in another direction, not wanting to disturb the equilibrium of a pleasant evening. For years I have asked friends and strangers alike what they think of wind turbines. If there is a dominant answer, it is often ambivalent, something like: "I don't like their intrusion into open space and the landscapes I love, but I recognize the good that they do, and that we have no other alternatives." The question of contrary attitudes seemed worthy of a more focused study.

A Gathering at Bellagio, Italy

In the late 1990s Paul Gipe and I decided we needed to talk more formally about the NIMBY effect. We wrote a grant proposal to the Rockefeller Foundation to bring together ten experts at Villa Serbelloni, the foundation's stunning retreat center at Bellagio on Lake Como. The foundation recognized the importance of such a discussion, allowing us to bring a team of four Americans and six Europeans to Bellagio for ten days. This international group, all involved with wind energy, crossed disciplines to include geography, engineering, landscape architecture, history, industrial design, the visual arts, and philosophy. Each participant wrote a paper discussing NIMBY from the standpoint of a particular discipline and

A common complaint is that wind farms industrialize the landscape. The barbed-wire fence, telephone poles and wires, and criss-crossing grid towers and wires with hundreds of wind turbines in the background make an uninviting scene. The Tesla substation at Altamont Pass is a good example of how a power landscape should *not* be built. (Author photo.)

location, providing the basis for invigorating and sometimes heated discussions. After the conference we published our work, the only book to my knowledge devoted to the issue of wind turbines and the NIMBY effect.[4]

The deliberations were wide ranging, but Gordon Brittan, a philosophy professor at Montana State University, took on the task of summarizing some of the ideas that occupied us. There were three areas of disagreement. The first was *perspective*. One group believed that wind farm developers must become more sensitive in mitigating the visual impacts of wind energy on the landscape, largely by more selective siting of the turbines and more involvement of the public in the process. Others believed that the answer to NIMBY was to reorient the way we think about wind energy, recognizing that unattractive turbines placed on revered landscapes is the price we must pay for our profligate use of energy.[5] Every electricity-producing method has externalities that are costly and often hidden. Perhaps annoying wind generators represent an aesthetic cost that can teach people how precious electricity is and encourage conservation. In other words, turbines are ugly—but the public produced the problem and must now live with it. Turbine retribution is the price we pay for a lavish electrical lifestyle.

The second main difference of opinion was in *aesthetic presuppositions.* Some agreed with the idea of beauty being in the eye of the beholder, a matter of individual taste. Others believed that some aesthetic judgments are objective and that there are universal standards that can be applied in decision making.

Four Problems

The third category may be described as various *problems.* Brittan mentioned four: "(1) the character of the technology, (2) its deployment in the landscape, (3) the system of ownership and control, and (4) the attitude of people to its increasing presence."[6] By the character of the technology we referred directly to the design of the modern wind turbine, questioning whether the Danish three-bladed turbine represented the culmination of technological advance. I and others argued for engineers to design and create alternatives, or at least to consider the aesthetic qualities of vertical turbines rather than the standard horizontal design. The more knowledgeable engineering participants, such as Paul Gipe and Danish designer Frode Birk Nielsen, thought the discussion was pointless, and there would be no "silver bullet" turbine that the public would find more acceptable. By 2009 Gipe and Nielsen have been proven correct, but some of us still feel that inventive engineers could produce reasonably efficient turbines that would lessen the NIMBY objection. The turbine need not be as efficient as the Danish three-bladed model. In some circumstances it would be sensible to offer a choice of turbines for a particular landscape. Whether that choice will ever be available is problematic. Yet there is no denying that many people find the basic turbine model large, intrusive, and ugly.

Size Matters

In regard to turbine dimensions, in 2000 we did not envision the increase in size. At that time a 500 kW to 1 MW turbine was large, and for a variety of reasons we believed they would not become much larger. We were wrong, of course. Today sweeping turbines of 1.5 MW to 3.6 MW are becoming standard, and the trend is to larger rather than smaller machines. They do not impact a landscape as much as dominate it. Turbines with blades reaching 400 feet into the air dwarf everything within sight. They command awe. On the West Coast we practically worship the coastal redwood trees. Some of the new turbines approach in height the tallest trees in the world, and our reaction is often the antithesis of worship. Their size makes it practically impossible to suggest that wind turbines can blend technology with nature. The water pumpers, the old icons of the American West, could harmonize with the natural environment, but there is no such symbiosis with the huge modern turbines. They are so *visible*. Engineers cannot hide or camouflage them nor build transparent or invisible turbines. The only thing they seem to be able to do is make the units bigger. Wind farm developers today have no answer to a distraught Vermont woman who asked: "What ever happened to 'small is beautiful'?"[7] She wanted windmills but felt they should be less conspicuous than the giant blades casting immense flicker shadows over her small state. Is there any chance that she might get relief? Not likely. The engineers and the economies of scale now dominate.

The second problem was *deployment*: how and where would we install these turbines? At Bellagio we stressed that local government and local people must be consulted from the beginning. Laurie Short, an English artist who bridges the gap between art and architecture and public perceptions of landscape, suggested that we must accept "technological fatalism"; that is, we really cannot affect decision making in the placement of wind turbine farms. We can only "change people's aesthetic perceptions." Although it appears that local input is hopeless, Short nevertheless stressed consultation and cooperation. Wind energy developers must realize the "important links among landscape, memory, and beauty in achieving a better quality of life."[8] This is a concept not always appreciated by wind developers, resulting in bitter feeling, often ultimately reaching the courts.

Laurie Short and also Karen Hammarlund, a Swedish social geographer, held that from the initial planning of a wind farm, the public must be informed. Hammarlund suggests that wind farm plans are often dominated by an "engaged elite," company engineers and executives driven by efficiency and profit, who have a clear idea of what they want but have little sense of local values. She believes that democratic consultation without too many preconceived notions (that is, flexible and open to change) is the best strategy. "Public acceptance is our best guarantee for a successful wind power development on land or sea."[9]

Closely connected to deployment is the *system of ownership and control.* In the United States, free enterprise is surely dominant. From California beginnings, wind energy manufacturers combined with developers and investment bankers to propel turbine growth on an amazing, ascending curve. They took advantage of lucrative federal and state subsidies and approached local government officials for property tax write-offs. They sold their power to the two giant investor-owned utilities in the state: Pacific Gas and Electric Company and Southern California Edison. Not much has changed since, except in scale. Today projects cost multiple millions, approaching billions of dollars, and Wall Street investment firms provide capital on a scale unimaginable to the small companies during the pioneer period.

However, there has been a shift in ownership and control. In the pioneer days such companies as PG&E and SCE were not interested in building or owning wind farms. They negotiated contracts for purchase of wind power, but ownership was far too risky and unnecessary. That has changed. One of the biggest wind energy producers is FPL, a subsidiary of Florida Power and Light, an investor-owned utility. Oil companies, with record profits to spend, also have been investing in wind installations. Wind is a good investment, and it softens the image of oil companies, so that they too, in small measure, can join the band of green brothers concerned with global warming and our energy future. But the public discerns that wind farms are no longer run by struggling environmental idealists, prospering one month, going broke the next. People today certainly believe in alternative energy, but they do not fool themselves that the first priority of wind farms owners is the environment. Owners talk the green game, but profit is their primary motive.

All this activity has meant that wind energy farms have proliferated across the nation, and so has the NIMBY effect. At Bellagio our group of ten believed that the NIMBY issue would diminish with new designs, better siting, and greater sensitivity to opposition. However, as more and more communities cope with the decision of whether to allow wind energy or keep it out, the NIMBY issue is alive and well from coast to coast. Demand for clean energy is on the rise. States apply pressure to meet their RPS goals, and the federal government sweetens the pot with subsidies. Most wind energy developers understand that the time for growth is now. Furthermore, larger global issues trump local concerns of impacted residents. In the give and take between wind farm developers and affected communities, the developers hold most of the aces.

Without question, wind energy represents the most benign way we can generate electricity. Yet this source faces determined opposition. Visual intrusion is at the heart of opposition, but there are other issues involved, one being the very emotional issue of avian mortality.

NIMBY and Avian Mortality

Perhaps the most historic NIMBY case occurred in 1987 when Zond Systems (which became Enron Wind, and finally GE Wind) leased 270 acres on Tejon Pass with the intent of erecting 458 turbines. It would be the first wind farm in Los Angeles County, a region sorely in need of clean energy. Immediately the Save the Mountain Committee formed in opposition, backed by the 200,000-acre Tejon Ranch. Real estate values dominated the ranch management's position. The directors wanted no turbine intrusion, for they intended to subdivide much of the ranch eventually. Zond president Jim Dehlsen offered compromises, and he enlisted the support of World Watch, Amory Lovins's Rocky Mountain Institute, Zero Population Growth, and Ralph Nader's group. But the Save the Mountain Committee was equally active, rallying the local Sierra Club and Audubon Society.

The showdown came on August 16, 1989, when the Los Angeles County Regional Planning Commission held a public hearing. Over two hundred people packed the meeting room. They heard Linda Blum of the Audubon Society as well as state wildlife officials testify that Zond had located the proposed wind farm in California condor habitat. This was a serious issue. State, federal, and private wildlife experts had invested much time and effort in a bid to save the condor from extinction. Wildlife officials considered any threat intolerable, no matter how slight. Although Zond promised to monitor radio signals from the condors continually, shutting down the turbines when the great birds entered the vicinity, it was to no avail.

Zond's chances came to an end when an elderly man took the podium representing the California State Racing Pigeon Organization. He lovingly described the beauty of his racing pigeons, their speed and grace, and his admiration for them. Then, in a dramatic peroration, he declared that if the Zond project went through, "our birds would look like they

Avian mortality is a serious problem for some wind farms. This red-tailed hawk (center) could lose its life to a spinning blade. Today wind companies have replaced lattice towers with tubular towers, making it impossible for raptors to use this high vantage. Yet in many regions, the death of migratory birds, bats, and raptors commands the attention of bird protection groups as well as wind farm operators. (Photo by Shawn Smallwood, NREL.)

went through a Cuisinart." It was the perfect sound bite. The image of chopped up pigeons and raptors, executed by turbine blades, was telling. The planning commission unanimously rejected the Zond proposal, and the Tejon wind farm idea died.[10]

The Tejon wind farm story underscores both the importance of bird mortality in any project and the power of local opinion. Although Zond did an admirable job of highlighting the macro-scale benefits of wind energy, they were no match for the micro-scale concerns of local opponents. Environmental writer Alston Chase castigated the local NIMBY reflex "dressed in the language of ecology." He believed that "the spirit of John Muir, Teddy Roosevelt and Rachel Carson is being co-opted by affluent practitioners of primitive chic more concerned with property values than with ecological sustainability."[11]

With regard to avian mortality, the thousands of turbines spread over the Altamont Pass hills of California have been most controversial. Dairy cattle still graze among the turbines, representing the traditional use of the land. It is good grass, an inviting habitat for field mice, voles, and various other raptor snacks. Golden eagles, red-tailed hawks, American kestrels and burrowing owls patrol the hills and rarely go hungry. However, the raptors did not adapt well to the new turbines, for developers ignored their needs and habits. Workers soon found numerous carcasses under the turbines. Blades and uninsulated wires were the primary culprits for the carnage. The unhappy situation persists. A recent Altamont Pass wind resource draft report estimates that between October 2005 and September 2007, the turbines dispatched 2,856 raptors and 8,765 nonraptors: a total of 11,621 deaths.[12] The Audubon Society and others have protested the killer turbines, but no solution has yet emerged.

In spite of numerous studies, there seems little that can be done at Altamont except to make apologies and occasionally shut down the turbines. The American Wind Energy Association (AWEA) recognizes that this situation is a real problem but considers, it largely limited to this area and not widespread. Unfortunately, media coverage about Altamont often gives the impression that all wind power projects have a significant effect on birds, despite overwhelming evidence to the contrary.[13]

Often the best defense is an offense, and that has been the strategy of the industry. Evidence shows that bird kills per megawatt of turbines nationwide number around one to six birds per year. If we use a figure of 25,000 MW of capacity and multiply that by four we have a bird mortality figure of 100,000. At a few sites, claims AWEA, no kills have ever been found. Industry spokespersons compare that statistic to annual bird kills by cats, 1 billion; buildings, 100 million, vehicles, 60 to 80 million; and pesticides, 67 million.[14]

The loss of raptors, other migratory birds, and bats may be the price that the environment must pay for energy sustainability. That has been the position of the Massachusetts Audubon Society in response to the proposed Cape Wind project. The society has acknowledged that birds die from the spinning blades but views the overall advantages of wind energy for the future as benefiting both people and birds. There are always trade-offs, gains, and losses—reactions for every action.

John W. Fitzpatrick, director of the Cornell Laboratory of Ornithology, has lent his expertise to the debate. On the choice between birds and wind energy, he says that "amid much hot air on both sides of this debate, research is beginning to shed light." No other wind farm has come close to the mortality figures for Altamont. He sees two reasons for the problems at Altamont: the lack of knowledge and concern in the 1980s regarding siting, and the high number of revolutions per minute of the old U.S. Windpower turbine blades, some still in use. Now, the situation is different. "Modern turbines rotate much more slowly than the original versions at Altamont," states Fitzpatrick, "acting more like ceiling fans than Cuisinarts and greatly reducing the potential for collisions."[15] Fitzpatrick also implies that many birds have the capacity to recognize the danger of the turbines and adjust. "Bird collisions appear to be rare," he says, and "radar tracking shows that migrating songbirds actually see and avoid wind turbines even at night"[16] Still, avian mortality remains an issue and wind developers acknowledge the problem. At the newly built Gulf Wind project in South Texas, developer Babcock and Brown spent three years tracking bird migrations. The company will use a precision radar system and shut down turbines during high-risk periods.[17]

In casual observations, Ken Starcher, head of the Alternative Energy Institute at West Texas A&M University, and I observed that the height of the huge turbines makes a difference. Standing amid them we tracked two red-tailed hawks working a field. They dipped and rose, but never more than thirty feet from the ground, while the low point of the sweep of the turbine blade was at least 60 feet. I was encouraged, believing that the habits of the hawks and the sweep of the turbine blades were not in conflict. The state of Wyoming is studying how turbines may affect sage grouse. Like the hawks, sage grouse fly close to the ground. However, they fear predators from above, such as hawks and eagles. Even though there would be no perching place for a raptor on a modern tubular tower, the sage grouse do not known that. Will they abandon their habitat if huge turbines arrive?

The bird issue continues to generate frustration and perplexity at Altamont and elsewhere. It will not go away. To alleviate the problem somewhat, the industry established the National Wind Coordinating Collaborative, which has published a siting handbook and avian site evaluation guidelines. They want to avoid any repeat of what one representative from Babcock and Brown called "a black eye for the industry." He reasoned that it had taught developers "the importance of doing these [avian] studies up front."[18]

The mortality is not limited to birds. In the East, significant bat kills have been associated with wind farms. People have a newfound respect for bats and their role in the environment. We do not want them to disappear or suffer a population decline. Some songbirds, such as warblers and thrushes, have also been affected. There is evidence that the raising of wind turbines on known bird migration routes down the spine of the Appalachian Mountains is an invitation to significant avian mortality.[19]

Perhaps the greatest threat to the growth of the industry is when an endangered species might be affected. Tom Stehn, whooping crane coordinator for the U.S. Fish and Wildlife Service, raised fears when he announced that "basically you can overlay the strongest, best areas for wind turbine development with the whooping crane migrations corridor."[20] The corridor runs from Texas northward through Oklahoma, Kansas, Nebraska, and the Dakotas to Canada. If T. Boone Pickens and other developers have their way, the corridor could contain thousands of turbines.

Naturally, a whooping crane crisis causes alarm. Laurie Jodziewicz of AWEA's siting policy committee has said such collisions "would be distressing for everyone." The 250 or so existing whoopers, like the California condor, are protected by the Endangered Species Act and the Migratory Bird Treaty. Nicholas Throckmorton, a spokesman for the Fish and Wildlife Service, states that birds and bats "have gotten used to flying around lots of things, but nowhere in the natural world is there a big spinning rotor."[21] Since the whoopers fly at 500 to 5,000 feet in their migration, one might think there would not be a problem. However, as Tom Stehn emphasized, they land every night, and it is during landing and taking off that turbine collisions are a risk. Surely the whooper issue will be high on the agenda of the Department of the Interior's newly formed Wind Turbine Advisory Committee.

Audible Intrusions

The other effective NIMBY blueprint is noise. Most urban dwellers would hardly ever hear a wind turbine over the normal ambient noise, but the countryside is a different matter. Depending on the weather, atmospheric conditions, topography, and the size and design of the wind turbine, noise can become a significant factor. I have been very close to large turbines in West Texas and Oklahoma and found the decibel levels low. The small, fast-spinning U.S. Windpower turbines on the Altamont Hills were much more intrusive. Most of the complaints are associated with noise at night. One Palm Springs resident who lives a mile away from the installation related that at night the turbines' constant thump-thump-thump was like his heart beating—a constant reminder of his mortality. Residents of the tiny town of Medicine Bow, Wyoming, could hear the air displacement noise of the Boeing MOD 2 and the Hamilton-Standard WTS-4 from over a mile away. Residents near the ESI model turbines at Altamont were constantly annoyed by a "high-pitched aerodynamic whizzing sound." Compounding these disturbances, the rotor created a thumping, helicopter-type noise each time a blade went behind the tower, a common problem with downwind turbines. Two residents, Darryl Mueller and John Soares, grumbled about the U.S. Windpower turbines, noting that although three-quarters of a mile away, they sounded as if they were "beating against my windows late at night" to the point that "it's almost unbearable."[22]

Across the country, county planning boards and supervisors are debating and enacting codes for wind farms. At a minimum a turbine must be 1,000 feet from homes; more often the requirement is at least half a mile. However, in many cases that is not enough, for noise travels far under certain atmospheric conditions. Opinions differ regarding the forty-six-turbine Michigan Wind Park in Huron County. Most nearby residents

Sometimes there is concern about the impact of the turbines on other wildlife as well as birds. However, these three elk seem unconcerned about the massive turbines towering above them. (Puget Sound Energy photo.)

agree with Rick Schmidt, who admits that the turbines make some noise, but "we are getting used to them." However, Jan Sageman is appalled by the din. She has no problem with the appearance of the turbines, but the noise can drive her to distraction: on cold nights with a light wind "you would have swore a train or jet was coming through the house."[23] Night time, when ambient noise is low, should be the testing time, for a wind turbine should not be allowed to invade a home and rob residents of their peace of mind.

Resolving noise issues is not easy. Low-frequency noise and vibrations can affect two people standing side by side differently. Richard James, a noise control engineer from E-Coustic Solutions, acknowledges that the way people respond to noise is variable. Sometimes we are conscious of a tumult of sound; at other times it is less obtrusive; and once a particular noise catches our attention, it can become not only annoying but an obsession. Ed Thompson of Meyersdale, Pennsylvania, lives with turbines a mile distant from his home. He reports:

> Prior to the wind turbines, these conditions [calm at ground level] would have meant a peaceful, quiet day or night. Instead, the turbines dominate the auditory scene. We have never gotten used to it. We put up with it, because we can do nothing about it. But, make no mistake, wind turbines do make noise and they will impact you if you live within a mile of them, no matter what the wind power companies tell you.[24]

Turbine noise can also influence property values. In Pekin, Illinois, Chicago real estate appraiser Michael McCann gave testimony that although proponents claimed wind turbines do not present an eyesore, he predicted a 20 to 30 percent drop in land value based on noise. One family is suing Gamesa Energy, a Spanish wind turbine company, for what they consider violations of noise levels under the ordinance agreement. Attorneys representing Todd and Jill Stull say that although billed as quiet, the turbines are loud. The excess noise is not constant, but when it occurs, Todd Stull must go to his basement to get any sleep.[25]

The only conclusion so far is that noise issues are difficult. Wind turbine noise affects individuals differently and decibel levels vary at different times of day and under various atmospheric conditions. Ed Thompson warns that "deception is at the heart of the message put forth by wind energy advocates." It may not be a fair accusation overall, but as in any business, there are those who would twist the truth to make a sale or sign a contract.

NIMBY as Competition

Electricity producers prefer a situation of shortage, essentially when demand exceeds supply. When wind energy developers enter a region where there is sufficient energy supply from another source, then they may face a Don Quixote of a different stripe. In Kansas, for example, coal-fired plants are dominant, but wind is plentiful. The coal industry views wind developers as an invading army, determined to conquer or cripple the status quo. When Sunflower Electric Power (certainly a misnomer) proposed two new 700 MW coal-fired plants, state regulators denied the permit based on CO_2 emissions. Political power came into play as the state legislature passed bills to overturn the ruling. At that point Governor Kathleen Sebelius vetoed the bills and the legislature narrowly sustained her decision.

Thus the governor's actions cleared the way for more wind energy development. When I traveled east on Interstate 70 in August 2008, new turbines were sprouting up in Kansas east of Hays and west of Salina. To my mind they are a marvelous sight. It is not likely the legislature will ever issue permits for the coal plants. But it was a difficult and costly battle. *Businessweek* reported: "So ferocious is this fight that Sunflower and its allies placed ads in newspapers suggesting that because Sebelius is against their coal project she's playing into the hands of Iranian President Mahmoud Ahmadinejad."[26] When large coal power companies play political hardball, NIMBY takes on a different air. Wind energy is clearly on the higher environmental and moral ground, but if coal interests combine with other wind opponents, they can place a serious barrier to the growth of wind installations.

While some objections to wind farms are clearly economically inspired and quite political in nature, no one can deny the legitimacy of many NIMBY responses. When the electrical power we want intrudes on the landscapes we love, there will be resistance, often passionate.

This is part of the democratic process. The vocal minority, if indeed it is a minority, has a legitimate right to weigh the pros and cons of wind development in the crucible of public opinion, in public hearings, and if necessary in our court system. Recently I visited the Arenal Volcano region of Costa Rica. At the western edge of Lake Arenal multiple wind turbines were at work. I asked two local people if their placement had been controversial. They said there had been no controversy, no hearing, and no public participation in the decision. The utility company put up the turbines with the blessing of the government. We do things a little differently.

A Concluding Thought

The power we want compromises the landscapes we love. Nature disallows the use of wind energy in much of the nation; should we disallow it where nature allows it? The American South has insufficient wind to offer much wind energy potential, a reality we cannot control. Except in the Atlantic coastal region, we will not see wind farms. In the American West wind is abundant. Sensible people will ask why opponents resist wind power development where it is so practical. As we face an increasing need to mitigate the consequences of our addiction to energy, local government and wind energy companies should do their very best to be attentive to those people affected by the turbines. In many cases, mitigation is the answer. However, in the final decision, personal concerns must not prevail. Global issues, and indeed our survival as a species, must win out.

Implied here is that whether we like it or not, we may have to sacrifice some of our natural landscapes for landscapes of renewable power. New York wind project manager James McAndrews reminds us that wind turbines present a very human-centered issue: "The trees, the deer, the birds care not if they see a wind turbine."[27]

McAndrews reminds us to look at the broader picture. So does the perceptive Maine environmentalist Bill Mckibbon, scholar in residence at Middlebury College and author of *The End of Nature*. He fully understands the paradox we confront. Speaking nostalgically of his boyhood days in New York's Adirondacks, he noted with remorse the recent rise of wind turbines on a ridge in the landscape of his youth. For Mckibbon the clock is ticking. He realizes that the world is warming, and unless we take steps to curb our energy appetite, the world "will change in ways we cannot begin to imagine." We have to change our perceptions and reconfigure our memory. After considerable reflection, Mckibbon can now "glance up at the turbine turning slowly in the wind and see it not as an ugly affront, but as a symbol of our willingness to take a small responsibility for the impact of our lives: To see it as something beautiful. The breeze made beautiful. The future made possible."[28]

CHAPTER 7

Addressing NIMBY

If a wind developer wants to get the job done, he must consult
with and consider the opinions of the "social landscape": that is, all
people who will be affected by change.
—Karin Hammarlund, social geographer, Sweden

All energy has a cost. . . . In the case of wind energy, that cost is the
destruction of the rural landscape on a national level.
—Bill Dolson, landscape artist, El Valle, New Mexico

If local landowners and the community oppose a developer's wind farm
project, what can an entrepreneur do? There are steps to take, but none offers any guarantee. No one can provide wind developers with easy answers
or formulas to overcome visual or auditory objections. They do not exist.
We all differ in how we react to our surroundings, natural or disturbed,
suggests geographer Yi-Fu Tuan.[1]

The American landscape is strewn with the results of our industrialization, most of which we scarcely notice. As Brian Hayes illustrates, over the
years we have learned to live with power plants, transmission towers, grain
elevators, oil pump jacks, communications towers, dams, canals, center-
pivot irrigation rigs, and legions of fences and utility poles.[2] Outside wilderness areas, it is difficult to view a landscape devoid of a human imprint.

Most of these intrusions are different from wind turbines in that they are stationary. Not only are wind turbines today huge, but they turn, annoying some but pleasing others. We need to understand the basis for opposition to the proliferation of wind turbines. Sometimes it is easily identified as economic. Either opponents are not getting enough lease income, or their neighbors are getting too much.

Cultural Roots of NIMBY

We will return to the economic problem after identifying the more subtle, yet nevertheless important, reasons for opposition. First, let us consider a sense of place. Our attraction to places is connected to *memory*, a powerful driver. People who have memories attached to a landscape do not want transformations. We all have special places—a park, a canyon, a vista, an area, a rural or urban setting to which we return regularly, if only in memory. The possibility that a special place may be altered by multiple wind turbines triggers determined opposition. Someone who rises at a public hearing and yells, "I don't want any damn wind turbines. Get out of town!" is operating on more than mere economic concerns. Robert Thayer, a geographer who was the first to study the NIMBY response to wind turbines, noted that at Altamont Pass opposition was strongest among those living close to the area.[3] This should come as no surprise. Familiarity with a place generates attachment and love. With love comes a sense of stewardship and a determination to protect the land as it is. Wind turbines can be anathema to that purpose, for they represent change.

Another subtle factor leads us all to question the dramatic transformation evident with a wind farm. Humans seem to have a cultural attachment to natural and pastoral landscapes. Even as we accept the industrial landscape of today, we long for the more natural world of the English countryside—winding roads, curving fences, wooded hills, hedgerows, and green pastures. If not the English model, we long for the simpler world of nineteenth-century America: a time when farming and rural life were the norm, and when the rising and setting of the sun determined our activities. Urban people jump in cars and head out of town for many reasons, among them to relocate in time, evoke a simpler past, and seek the stability or renewal that nature offers. We are not pleased when that ideal landscape has disappeared. Thayer calls this "landscape guilt."[4] Historians are more inclined to call it a form of anti-modernism—reluctance to accept the complexity of the modern world and a desire to return to simplicity represented by unencumbered landscapes. Clearly many Americans feel at least ambivalent about how technological development has impacted, indeed dominated, the land and nature.[5]

Wind developers tend to brush aside such subtle, often nebulous objections as mere intellectual foolishness. Yet impressions motivate action, and thus perceptions must be countered with ideas, not with legions of lawyers. Today, the proponents of wind power are much more adept at expressing ideas than they once were, often countering objections by proclaiming a higher purpose than mere business. Wind developers can capture the high ground of ideas because they are generating a product that can help save us from ourselves. Science can no longer be disputed. The globe is warming, and while the likely results are not altogether clear, they will not be pleasant. The scientific community has the evidence, and wind developers hold one kind of solution. They can claim with impunity that they are addressing a portion of a world problem. They can claim to be thinking globally and acting locally. This is a powerful weapon, and few opponents of wind turbines are able to combat it successfully.

Economic Issues

Do developers embrace the advantage of the moral high ground? More commonly they win over landowners by emphasizing economic gain. To generalize, companies are typically dominated by hard-headed executives and attorneys who fully expect to bowl over the opposition with money and relentless talk. In the United States developers usually offer landowner contracts stipulating a royalty for each kilowatt-hour produced per turbine per year. Contracts are not discussed publicly, although that is changing as landowners become more informed and sample contracts have become available.[6]

Once enough individual landowners have signed leases, the developer feels free to move ahead with the permitting process and the project. Left out in the cold are the adjacent landowners. There is no monetary reward for them; all they get are the turbines' negative effects. Depending on the distance, abandoned landowners suffer visual and noise impacts. They lose the natural attractiveness of the land, particularly the viewshed. Often the view, the quiet, and the natural beauty influenced the family's decision to purchase the land. Now it is gone. For them wind turbines represent an actual "taking" of the enjoyment and value of a family's land. They have involuntarily relinquished a right. The fiercest opposition comes from those who feel aggrieved from such injustice.

Is there another way? I believe there is a way that will help defuse the NIMBY response. It is called fairness, and it features community cooperation rather than competition. If they are to be fair, developers must be open, revealing their plans and the conditions of landowner leases. Nothing can be more destructive to community than secrecy. Initially, covert lease contracts seem attractive to wind developers. However, over the long run as disparities are inevitably revealed, hard feelings will be the harvest. Why not be open? Karin Hammarlund, a Swedish social geographer, has defused a number of local NIMBY controversies. Her advice is it is usually a mistake to allow an "engaged elite," local people of influence and/or opinion leaders, to dominate the dialogue. "If a wind developer wants to get the job done," suggests Hammarlund, "he must consult with and consider the opinions of the 'social landscape': that is, all the people who will be affected by change."[7] In short, the developer must consult the whole community, and without set plans or conditions. Probably we should alter the acronym NIMBY to NIEBY: Not in Everybody's Backyard.

Environmental Justice

Another NIMBY question that deserves attention has to do with the upper class. Environmental justice is a concept normally invoked to address unfair treatment of the lower class, but this interpretation can be turned on its head. Sometimes the power elites do not want environmental change, even if it is beneficial. When John Passacantando of Greenpeace debated Jerry Taylor of the Cato Institute regarding the Cape Wind project, John made it crystal clear that the opposition arguments were not based on avian mortality or fishing rights. Opposition was organized and paid for by the wealthy, the privileged of the Cape Cod region, who opposed any modification to their window gaze out over Nantucket Sound.

Anyone who peruses the book *Cape Wind* must come away with this conclusion. As earlier mentioned, the subtitle, *Money, Celebrity, Class, Politics and the Battle for Our Energy Future on Nantucket Sound*, alerts the reader to the theme.[8] I want to make the case for a different interpretation of environment justice. In this instance environmental justice does not involve the powerlessness of the poor to prevent contamination of their world. We are not referring to Hispanic farm workers' exposure to toxic pesticides, or Navajo workers' contact with coal dust and radioactive uranium tailings, or the tendency of sanitation companies to place dumps near poor communities. My contention is that environmental justice should insist on polluters taking responsibility for their contamination. Specifically, environmental justice would call for wind turbines to be built directly in front of the viewshed of the Mellons, DuPonts, Cronkites, Kennedys, and all the other wealthy neighbors who frequent the Oyster Harbor Club. These families with mansions, servants, and three-car garages are consuming a disproportionate amount of electrical power. Their carbon footprint is large. They may profess to be environmentalists and contribute money and buy carbon trade-offs, but that is not enough. Environmental justice suggests they should pay for their profligate consumption of energy.

Robert Kennedy, Jr., for example, has been an active environmentalist. In the May 2008 issue of *Vanity Fair*, Kennedy wrote a wonderful manifesto on the dramatic, and expensive, steps the nation must take to combat global warming. But ironically, Kennedy thinks globally while failing to act locally. He and the late patriarch Ted Kennedy have been defiant opponents of the Cape Wind project. You cannot fight a war against global warming without some sacrifice; that resembles the moneyed class during the American Civil War sending a servant to do their fighting (and dying). It is like the affluent of Santa Barbara opposing oil platform drilling off the coast. Environmental justice calls for the wealthy to make *direct* sacrifices. It is time we all shared environmental realities and the inconvenience they cause, and least of all should the affluent receive a pardon.

Wind turbines in our backyards remind us that electricity has a cost, and it must come from somewhere. We have not found a totally benign way to generate power, and in spite of the work of brilliant engineers, we are not likely to do so. Wind turbines, as geographer Martin Pasqualetti reminds us, are a visual price we pay for profligate consumption.[9] There is a kind of justice in a constant, annoying visual reminder of the need for conservation no matter what your economic circumstance. We need to turn off the lights and turn on the wind generators.

The Case for NIMBY or NIEBY (Not in Everybody's Backyard)

It is illuminating to look at some of the individuals and families who are caught in the vortex of the spinning blades and must endure them. In many cases they have been trampled by local or state government and neglected by the company that erects the turbines. Like Dale Rankin of Tuscola, Texas, they are expected to make landscape sacrifices, finding solace in the global vision, not harping on their lost view or the end of peaceful nights. They are asked to remember the benefits of an energy source that emits neither heat nor pollution and that is without major externalities, such as the cost of providing fuel to a generating station or the immense tab to process nuclear waste. No matter how admirable this is, should a few people pay the price for benefits to the many? Should rural regions lose the amenities and psychological comforts of living there to serve the city? Should metropolitan areas enjoy abundant electricity while rural people forfeit the very qualities that took them to the countryside in the first place? The macro-scale benefits of wind energy seldom impress local opponents, who have micro-scale concerns. The turbines' benefits are hardly palpable to impacted residents, whereas the visual impact is a constant reminder of the loss of a cherished landscape.

The stories of people and communities affected by the giant turbines are legion. Sue Sliwinski of New York decided to spend nine days covering three thousand miles and visiting seven wind farms. Taking video, still shots, and notes and conducting interviews, she found that all the impacts denied by wind farm developers did indeed exist. "Lovely rural communities," she noted, "are being turned into industrial freak shows." She told of a woman who, when outside, learned to block the turbines' sound. A man in Lincoln County, Wisconsin, suffered severe "stray voltage" on his rocky farm. His neighbor experienced the same problem, but the neighbor had ten to fifteen turbines on his land and chose not to jeopardize his income by complaining. By this writing he has probably sold his farm and moved to a more hospitable location. Some have sought redress in the court system. In Illinois, a farmer and dedicated wind turbine opponent has spent over $250,000 in court battles. He failed to get redress, but that does not make his loss any less real. He finally bought a cabin in the woods, seven miles from his home.

In the Mendota Hills wind farm site in Illinois, Sliwinski found a near deserted community. Lawns were mowed, but every window and door was closed. She came back at night to find only one light burning in the houses and no movement whatsoever. In another community children were terrified by the night noise, especially on rainy nights, and could not fall asleep. Nearby an older woman cried as she told how the town where she was born and raised was now ruined.[10]

People write letters to their local newspapers expressing outrage. Julie Gullickson of Brownsville, Wisconsin, vented her feelings like this: "Now all I hear when I go outside to my once quiet yard is that thumping swishing noise created by those stupid wind turbines. Not to mention how ugly those things are with their stupid red lights on top of them." She continued: "This is not a God-made creation, they are man-made pieces of junk that are creating a lot of noise in my yard! I want something done about this!"[11] She complained that she now lives in an industrial park. Her vehement opposition, although unsophisticated, is representative of the views of many ex-urbanites who believe they own the working landscape.

Much of the NIMBY reaction to proposed wind farms has been expressed at public hearings. People feel intensely, and these hearings can be long and volatile. In Illinois in early April 2008, Tazewell County's Zoning Board of Appeals held a public meeting on the Horizon Wind Energy company's special permit application to erect thirty-eight turbines in the county. Citizen after citizen spoke out against the project. The objections were varied, but perhaps the most common was the belief that once the turbines went up, property values would go down. After three hours of testimony the last few speakers were forced to limit their comments. Clearly it was an emotional meeting. The board scheduled another hearing.[12]

Perhaps the most grievous situation is when a wind farm tears a rural community or family apart. In New York the Maple Ridge Wind Farm consists of 195 Vestas 1.65 MW turbines, producing 320 MW of capacity. It is jointly owned by PPM Energy of Portland, Oregon, and Horizon Wind Energy of Houston, Texas. The two companies' websites claim the landowners are more than satisfied and have nothing but praise for the income and prosperity the turbines have brought to a struggling farm area. One resident, Carl Stone, even wrote a poem in praise of wind power.[13]

Yet all is not bliss. The NIMBY effect is alive and well. Arleigh Rice, a town supervisor for Lowville, New York, one of the towns that hosts the wind installation, laments that it "has caused friction, family against family or neighbor against neighbor." The Yancey family, having farmed the land for at least fifty years, is disintegrating as a result of the wind turbines splitting the father and daughter against three sons. Patriarch Ed Yancey decided to allow seven turbines on his land. Daughter Virginia sided with her father's decision, but sons John, Herb, and Gordon opposed the idea. The senior Yancey could not resist the approximately $45,000 a year he receives in lease monies. Son Herb's opposition is based on the fragmentation of the farm, making working the land "a nightmare." Gordon regrets the loss of his view of the Adirondacks, now obstructed by towers and turning blades. Most seriously impacted has been John. One of his father's turbines sits just across the road from his house. When the wind is right, the sound penetrates his bedroom. "I don't sleep," he complains.[14]

Ed Yancey's dilemma pitted money against family considerations. In retrospect, the elder should have consulted his sons and worked out a fair compromise. He did not, and son John may move away from the family compound forever.[15]

Are the developers of the Maple Ridge Wind Farm responsible in any way for this family schism? Under the rules of the game, they are not. Neither Horizon nor PPM Energy had any obligation. The wind energy business throughout the nation operates on the free enterprise system without any responsibility to ensure community or family agreement. One might assume that an Environmental Impact Statement would address such socioeconomic factors. But the Yancey family conflict fell under the radar screen—one of the victims of wind energy development. Again, should the wind companies shoulder the blame? I believe they should. Good corporate citizens must identify potential problems and take action, and that action should precede final placement of the wind turbines.

Conflicts between landowners and wind developers have a commonality throughout the nation. Upstate New York and New Mexico are geographically and culturally separate, and yet the two states are both disrupted by wind energy conflicts. In New Mexico battles are brewing over intrusive wind farms. Unless a wind farm is over 300 MW in size, of which there are none in New Mexico at present, no state regulations apply, and most counties have few ordinances to regulate wind energy and protect rural residents. Fights become local, and as such, they are intense. Rural residents question whether the turbines "have to turn ridgelines into scenes from science fiction, or can they be nearly as effective in less-obtrusive nearby locations?" It is an old question, and occasionally the answer can bridge the chasm between the company's electrical engineer and the local residents. The most optimal ridge need not be developed at the expense of residents' rights to the enjoyment of their property. In such a state as New Mexico, solar panels, which have been greatly improved in efficiency and reduced in cost, may be a better renewable energy option than the wind turbines. Bill Dolson, a landscape artist living in El Valle, New Mexico, notes that renewable energy is not free. "All energy has a cost. In the case

Although the placement of turbines on this section of the Maple Ridge Wind Farm in New York looks relatively idyllic, there is always opposition to the appearance of wind turbines on private and scenic land. Note that one farmhouse is visible with three turbines in close proximity—too close. (NREL photo.)

of wind energy, that cost is the destruction of the rural landscape on a national level."[16]

Mitigating Opposition

From these examples we can see that there is justifiable opposition to turbines across the land. Huge wind turbines are impacting the national landscape, and no one should minimize that fact. Companies must be responsible. Their public relations people should participate in the local governmental process. They need to listen to the concerns of locals, and they must *anticipate* the effect of their project once completed. They must be good neighbors. One local landowner, pleased with two turbines on his land, believed that those who had opposed the windmills would soon get over their concern. "Does anybody pay any attention to electric light poles on the side of the road?" he queried. "That's exactly what windmills are going to be like in a year." Such a preposterous analogy should be challenged by company public relations officials. Contrasting a stationary forty-foot utility pole with a *moving* turbine reaching above 300 feet in the air is inappropriate. Companies allowing such flagrant interpretations to pass unchallenged are gaining a short-term victory at the expense of losing the goodwill of much of the community.

Should we expect wind developers to be honest with the public? Should we expect these developers to be forthright with lease holders and adjacent landowners? Again a doctrine of fairness should prevail, but often it does not. Companies that operate in regions with little regulation and oversight often drive a hard bargain, but it will backfire in the long run.

The NIMBY response to turbine development is often led by adjoining landowners who have no turbines on their land but suffer the visual effects of their neighbors' turbines. They have nothing to gain and everything to lose, like Dale Rankin. As already described, the Horse Hollow Wind Farm and its owners, Florida Power and Light, crushed Rankin with their lawyers when fairness and reason could have ameliorated the situation. With 47,000 acres under lease, and 291 GE 1.5 MW and 130 Siemens 2.3 MW turbines in place, the company could well have compromised on the siting of two turbines. But they did not.

Another Way: The Fairness Doctrine

The problem was that Dale and Stephanie Rankin had no legal right to stop the turbines, and FPL attorneys had no interest in what I call the fairness doctrine. A policy whereby contiguous landowners vigorously protest construction because they get nothing is guaranteed to foster divisiveness, one landowner pitted against another. Who can blame the adjacent owners? They suffer visual and auditory abuse without any economic benefits. They see neighbors getting rich on wind royalties while they struggle. It does not seem fair, and indeed it is not. The fairness doctrine is nothing more than responsible business ethics.

Is there another way? Previously, I described the Danish commitment to community wind energy efforts. It is worth looking also at some other European countries, for if the answer to overcoming NIMBY is fairness, Germany, France, and Spain can all give us some ideas. First, because these countries have national feed laws, the price a landowner receives is not negotiable; it is set. Therefore there is no squabbling in Germany about contracts and price. "Everybody knows the ground rules," explains Paul Gipe, "and the price paid is the same for everyone, whether for a pig farmer or a wind company." In effect, the playing field is level.

Establishing a set price for wholesale electricity resolves one irritant. We also need to look at Europe with regard to landowner royalties. In Denmark many of the turbines are owned by small farm communities, and both the costs and the profits are borne by all. More to the point, in France adjacent landowners receive 30 percent of the royalties paid. The local planning authority believes that such owners, who must live with the turbines but do not have one on their property, should received some compensation. One project developer, Erélia, moved some turbines slightly from their ideal locations to neighborhood properties. Payments are also made to landowners for passage of roads and cables. Furthermore, the company provided payment information to *all* property owners, in marked contrast to the secrecy often prevalent in the United States.[17]

Policies in Germany likewise respect the fairness of sharing royalties. Typically land leases receive 5 percent of the gross revenues from each wind turbine. One percent goes to the nearest villages. Of the remaining 4 percent, half is paid to the landowner and the other half is distributed among surrounding landowners.[18] Such financial sharing in the proceeds eliminates the all-or-nothing scenario.

In Canada's Prince Edward Island people have borrowed a page from the German system. On the eastern end of the island sits a 30 MW wind farm of Vestas V90s, operated by the Prince Edward Island Energy Corporation. The area is windy and the Prince Edward Island Minister of Energy,

Jamie Ballam, expects that each turbine will produce about $20,000 (Canadian) in royalties. But contrary to the American system, the royalties will be divided into three zones. The landowner will receive 70 percent, or $14,000 per turbine. Those landowners within 100 meters (concentric circle) will receive 20 percent, about $4,000. Those within 300 meters will receive 10 percent, or $2,000.[19] No one would claim that such a pooling arrangement will resolve all inequities, bad feeling, and the other negative aspects of turbines on the landscape, but at least the effort of European countries and Prince Edward Island officials is a step toward fairness and addressing the NIMBY response.

Build Them Where They Are Wanted

In the final analysis we can best address the NIMBY response by building wind turbines where they are wanted. Unlike Europe, our nation has land. There are vast areas of the United States that have excellent wind resources and welcome the wind turbines. I am thinking specifically of the Great Plains: the vast region west of the ninety-eighth meridian and east of the Rocky Mountains. Included within the area are West Texas, the western lands of Oklahoma, Kansas, and Nebraska, and most of North and South Dakota. Portions of New Mexico, Colorado, Wyoming, and Montana would be included as well. It is a vast and windy area, perhaps one-fifth the land surface of the United States.

North and South Dakota were portrayed as "the Saudi Arabia of wind energy," a description to which Texas and Oklahoma oil millionaire T. Boone Pickens would subscribe in the media blitz in the summer of 2008, mainly on television. He favors expanded natural gas use and erecting thousands of wind turbines to solve our energy crisis and oil dependency. He would start the project on his own land north of Amarillo and, if offered the proper incentives, move north all the way to the Canadian line. Pickens is obviously a man who thinks big, and the downturn in the economy has temporarily put his ambition on hold. But in our current political climate favoring energy and jobs, that could change overnight.

Would such a massive project raise insurmountable NIMBY objections? I do not believe so. Of course there would be special landscapes that should not be sacrificed, such as the Black Hills, Scott's Bluff, Tallgrass Prairie, and the Badlands, but much of this vast, essentially flat, windy land is well suited for energy production in this century. Since settlement, it has been a land dependent on wind. One hundred years ago, well over two million water-pumping windmills allowed ranchers to run cattle on these semiarid plains. Most windmills are gone now, abandoned for electric pumps, or more likely simply deserted by owners, a symbol of a disappearing rural America. However, while the machinery no longer works, the wind still blows.

In the late 1980s two land use planners from Rutgers University, Frank and Deborah Popper, observed the depopulation of the Great Plains. They noted that deaths vastly exceeded births, schools were closing, commodity prices were declining, and family farms were being sold, with their contents auctioned at depressing prices. They had no silver bullet for the economic decline. But the Poppers did suggest another way forward: turn the vast landscape into a "buffalo commons." Local people viewed returning the land to the federal government as a horrible solution. The Poppers became characterized as academic Eastern snobs. Residents roundly criticized the idea, and the Poppers were considered candidates for lynching. Although they may have been insensitive, the Poppers were correct; residents trying to make a living from the land have seen the wisdom of predictions of decline.

Placement of a wind farm on arid or semi-arid land generally faces little opposition. As in this West Texas photograph, often no one lives within sight, and those who may be impacted usually believe that the economic benefits outweigh any landscape loss. (NREL photo.)

In 1893 historian Frederick Jackson Turner announced that the American frontier no longer existed since most of the land in the American West land supported more than two people persons per square mile. That demographic figure has reversed itself on the Great Plains. Population estimates in North Dakota in 1998 show that thirty-five of fifty-three counties have reverted to less than two persons per square mile. Kansas has more frontier counties now than it did one hundred years ago. Only the regional hub cities have grown in population, as young people look for jobs and a more exciting lifestyle.[20]

What I am suggesting is that immense sections of the Great Plains require economic assistance, and wind turbines can help. The primary obstacle is getting the electricity to market. Settlers on the Great Plains were always aware of their extreme isolation from urban markets and amenities. Power production by great turbines is no different. Government support, willing wind farm developers, investors, and progressive utility companies will be necessary to end the electrical isolation of the Great Plains. Such cooperation will counter the economic decline of the region. Thousands of turbines built by Pickens or other developers will provide jobs in construction and maintenance, substantial income for landowners, and a new sense of hope for a marginal area. There will be controversy because a portion of our population abhors industrialization of the landscape. But aesthetic NIMBY arguments will be minimized by economic and environmental logic. Planners are fond of determining the land's highest and best use. Wind energy will be triumphant on the Great Plains.

To continue this logic, we should develop wind energy farms where they are wanted and where they do not overlap with other land use options. Cows and sheep live compatibly with the turbines. Conversely, wind developers should give serious consideration to *not* insisting on raising turbines where they are not wanted. The United States has an energy advantage over England and Europe: we have vast stretches of windy, largely vacant land, waiting to become part of our energy solution. We can hope the industry will adopt the attitude of Bob Gates, a Clipper Windpower vice president: "If people don't want it [wind development], we'll go someplace else."[21] Fortunately, the country can accommodate him.

CHAPTER 8

Small Turbines and Appropriate Technology

Because of new U.S. subsidy programs for purchasers of small wind turbines and new technology breakthroughs, the future prospects for cost reduction in small wind turbines are the best they have been in twenty years.

—Bergey Windpower, 2009

Thus far we have focused on the big guys: turbines with blades reaching four hundred feet long, connected to public utilities, generating megawatts of electricity, and attracting Wall Street investors. There is another category of wind turbines that might be labeled the Birkenstock set, except that this description fails to recognize the variety of individuals who have been generating their own electricity. Most could be considered environmentalists in the broad sense. Some belonged to communes. Others sought to get back to the land. Some were not interested in farming but wanted to challenge the direction of American technology, particularly nuclear energy. Some just wanted a little electricity for a remote cabin. Most were idealists who opposed electrical centralism and the bill from their local utility. A few built stand-alone systems because they had no choice. In truth, for over one hundred years individuals throughout the nation have bought or built small wind turbines to meet their energy needs. The development of these small turbines is outlined in chapter 1, and we now return to them with emphasis on their cultural and lifestyle implications. I do believe the small turbines and their owners have a place.

Recall the bold peasant Herbert putting up a post-mill in twelfth-century England in defiance of Abbot Samson, who feared competition for his watermill. Brave Herbert and his sons were forced to dismantle their windmill, but not before Herbert uttered that "the free benefit of the wind ought not to be denied to any man." It is a significant statement because, in contrast to other forms of energy, wind is such that neither a powerful abbot nor any corporation or nation can control this energy source. Energy companies can lease or buy land, water, or resources like oil and coal, but they cannot lease or buy wind any more than they can buy the sunshine or the currents of the oceans. The wind is free and the price of this fuel never changes. It is available to any person who has a patch of land to put up an apparatus to capture it.

Such an egalitarian form of energy was bound to attract a large following. The estimated 6 million water-pumping windmills once at work throughout the American nation are evidence of how a simple intermediate technology machine could improve the lives of millions of western Americans. And although they probably knew nothing of Herbert, they fully understood the free benefit of the wind and how their little water pumpers lifted the drudgery of farm and ranch work from overburdened shoulders.

Today of course, the turbines are no longer little, and they produce electricity, a product that has revolutionized the way we live. From the 1880s, inventors worked to marry the power of the wind with electricity to serve the practical needs of homes, farms, or small companies. Contrary to what we might think, the early history of wind-electric energy centers on individual, stand-alone systems. A central grid system still does not exist in many areas of the world, particularly in developing nations. In some places the most practical way to offer electricity is through wind power or a hybrid system. Wind turbine–diesel systems have worked well, but small, remote villages may have little hard currency for fossil fuels. Increasingly, wind turbines are being matched with solar generation. Until recently solar panels were too expensive to be useful in third world sites, but that is changing. One prominent wind turbine manufacturer believes that often the lowest cost system is a hybrid that combines wind, photovoltaics, and diesel power.[1] When the sun is not shining and the wind is not blowing, users can fall back on diesel. Whatever the off-grid electrical mix, there is still a place for the wind turbine.

The evolution of wind energy has moved from the iconic water-pumping windmill to the electricity-producing turbine, both seen together at Kimball, Nebraska. Can we accept the new intrusion on the vast landscape? If we can learn to appreciate what these turbines do, perhaps they will become acceptable additions to our vast viewscape. (NREL photo.)

Connecting to the Grid

In the United States today most larger home turbines are connected to the grid, while the smaller varieties are off-grid. Statistics gathered by the American Wind Energy Association indicate that the 2008 sales of small turbines totaled 9,092, of which 7,800 were off-grid. In terms of kilowatt-hour production, the total capacity of the few larger home units was 5,720 kWh, while the 7,800 smaller turbines' capacity was 4,017 kWh.[2] Obviously, off-grid wind turbines are still popular and are being used in a variety of ways, mostly in situations where a limited amount of power is required. The larger residential turbines interconnected with the grid are taking advantage of an option unavailable until 1978, the year Congress passed PURPA, mandating that owners of individual turbines had the right to hook up to the utility company grid, both to buy and to sell power.

The option was not immediately available, since utility companies nationwide resisted buying electricity from the annoying small wind turbine owners. When forced, utility managers offered one-sided contracts that amounted to roadblocks. Families who had invested thousands of dollars in a turbine found that the utility company discounted their independently generated power, but they still had to pay full price for utility-provided power. One family in Casper, Wyoming, threatened to take their case to the local television station if their utility did not stop being obstructionist. Not wanting negative publicity, the utility fell into line; the television threat paid off. Utilities could not deny turbine owners their right under PURPA, but they could hand out plenty of grief through legal loopholes, price differentiation, and special equipment costs that had to be borne by the small power producers (the official government name for home power generators). Even recently, some rural Kansas cooperatives were charging from 12 to 19 cents per kWh, and yet their buy-back rate was only 1.3 cents per kWh, about 7 percent of the retail price![3] In effect, the utility companies overreacted to the annoying gnats that threatened their monopoly, an exclusivity hammered out in the time of Franklin D. Roosevelt and one that had served them very well.

A few utilities saw the wisdom of accepting the small power producers and treating them as equals. A spokesman for Ohio Edison explained that his company "will operate in parallel with customers provided they have the right equipment." Power firms in Alabama, Connecticut, Iowa, Maine, Michigan, Minnesota, Rhode Island, and Virginia took similar positions.[4] Just what the Ohio Edison spokesman meant when he stated that his company would "operate in parallel" is uncertain, but we can assume that excess energy produced by the wind generator would spin the existing home meter backward, providing customers full retail value for the electricity their turbines produce. This is called net metering, and for small power producers it is a crucial concept if the wind turbine is to pay for itself. If individuals have to sell their product at a much lower price than they pay for the same product, it is difficult for a home wind turbine to be much more than an idealistic environmental luxury.

Several other advantages apply to net metering. The wind turbine owner need not purchase another meter to measure electricity inserted into the grid. A standard kilowatt meter accurately registers the flow of electricity in either direction. Thus with net metering, the meter runs forward when the home draws utility electricity but backward when the turbine is producing more electricity than the home can use. The homeowner is billed for the net amount of utility energy used or, if it has been a good month for wind, is given credit for the excess energy delivered into the grid system. Among the advantages of net metering, even for the utility company, is its simplicity: it entails no complicated paperwork adding or subtracting kilowatt-hours. Approximately twenty-eight states now have some form of net metering legislation on the books. In other states, potential wind turbine owners must consider carefully the economic consequences of dependency on the whims of the utility company.

Government Policy

There is certainly a strong urge in many Americans to cut the power lines, declare our independence, and get off the grid. Forget the net metering and forget utility bills! Psychologically satisfying as this might be, it is probably not a good idea. Reliable power all day every day with an off-grid system requires batteries and also a backup generator. The wind turbine savings could be eaten up by the cost of batteries and fuel for a diesel generator. Furthermore, certain state and federal subsidies would not be available for an off-grid system.

California leads the states in encouraging small turbine support. For instance, a small producer can actually store or bank excess electricity for use at a later time. Energy banking is done over a twelve-month period so that electricity generated during windy months can reduce the electric bill during low-wind months. Only a few other states provide for net metering on an annual basis.[5]

From the pioneer period on, California offered small wind systems exceptional support. In the early 1980s, if you wanted to install a $5,000 solar heater for a swimming pool, you could write off 50 percent ($2,500) as a tax credit. According to Ty Cashman, who worked in the Office of

Appropriate Technology, getting a credit for wind energy was more difficult. But finally the legislature enacted a 50 percent tax credit for wind turbines.[6] These were halcyon days for wind turbines, both commercial and home units, but it all ended in 1985. State subsides ran out, and the California legislature chose not to renew them.

However, the state is now once again in the forefront after renewing its commitment to small wind turbine purchasers. In 2000 the state legislature granted individuals who installed qualified small wind turbines a rebate of up to 50 percent of the cost. This represents a remarkably generous rebate, which the legislature extended to 2012.[7] Qualified turbines must be no larger than 10 kW, have a five-year warranty, and be installed by a licensed contractor or electrician. The California Energy Commission also requires that a turbine owner must make a "reservation," since rebates are on a first-come, first-served basis until the available funding is exhausted.[8] Thus one must pay attention to the rules, but the rewards are great. For instance, the Bergey 10 kW Excel-S system is certified by the California Energy Commission and qualifies for a rebate of $21,450. With this incentive, the company estimates the payback for the turbine to be six to seven years.[9]

Incentives

Any study of the American electrical complex will reveal that the federal government has largely shaped how we produce and consume energy. On the national level, the American Recovery and Reinvestment Act of 2009 will surely be a boost to the small turbine business. It allows for a 30 percent tax credit for the total cost of a unit. For instance, the average installed cost of a Skystream 3.7 system, manufactured by Southwest Windpower, is $14,000, thus the incentive would be $4,200.[10] This incentive is not as lucrative as California's, but it certainly makes a difference. Depending on where the purchaser lives, there may be local tax incentives. Furthermore, if the buyer cannot take full advantage of the credit in one year, it may be spread over two. This wind energy stimulus is essentially identical to that given for commercial size turbines and will continue until December 31, 2016. With such incentives the growth of the big, commercial turbines will continue, and there is no reason that the small power units will not follow along, adding their more limited effort to bring us to energy sustainability.

With both federal and state incentives in place, small turbines are becoming more popular. Here a Skystream 3.7 turbine combines with solar roof panels. Assuming that the homeowner can take advantage of a net metering system, the electrical bill will be small, and at certain months of the year the utility company may be paying the homeowner. (NREL photo.)

Small Wind: A Dubious Past

Thus the small home power units are making a comeback. Some of this resurrection recognizes that the tiny wind turbines for residential use have advantages. The possibility of bypassing the traditional model of big, centralized stations is attractive. These transmission systems (the grid) have become expensive headaches. Why not provide electricity on a home to home basis? Why not generate electricity where it is consumed? The stimulus package will not change the delivery system, but it does signal that huge central power stations and grid systems are not the only way to provide electricity.

Although subtle, government policy has become more tolerant of decentralization. One is reminded of 1940 and the farmers of Dickinson and Harvey counties, Kansas, who surprised the canvassers from the state university when their survey showed that "there seemed to be no great desire for electrical service [central power] on the part of many farmers." The researchers found that 165 independent stand-alone units operated in Harvey County. Dickinson proved similar, where 215 wind or gasoline plants provided electrical power for family farms. The survey concluded that "complete electrification of the farms of Harvey or Dickinson County does not appear to be practicable under present circumstances."[11] The evidence from this report and other accounts is that rural people were far from being unanimous regarding central power. Today is not altogether different. Anyone with a few acres of land can think about going semi-independent by establishing a hybrid system of central and decentralized power.

Most of the farmers of Dickinson and Harvey counties soon succumbed to the lure of REA central power. Government policy gave no encouragement to stand-alone systems. Those who resisted central power were branded stubborn old fools. I met one of those stubborn fools in West Texas in the mid-1980s. His name was Joe Spinhirne, and when I visited him in 1985 he was still generating his own power from a Jacobs turbine he had bought for $100. While his neighbor ranchers hooked up to the REA cooperative, he bought their discarded 32 volt appliances for a song. When I met him he was laughing at his neighbors' escalating REA power bills.[12] Spinhirne was just one rather obstinate holdout, but there have been more organized protests. In Minnesota, the once-revered REA suffered sabotage of their electrical towers by the "bolt weevils," a group of radicals who opposed power transmission and centralization. These activist farmers were willing to face fines and prison sentences for their civil disobedience.[13]

My point, however, is to underscore the federal government's changed attitude toward stand-alone or hybrid electrical systems. The nation will never change from centralized to decentralized power. However, through subsidies and tax credits, public policy makers have acknowledged that the small systems do not represent a dangerous radical fringe and that the owners are not the anarchists of the energy world. We can have some diversity. There are probably two primary reasons why people put up a small wind turbine: first, to save money and get free of the despised utility company; and second, to do the earth a good turn by minimizing one's global warming footprint. Neither is imperative or subversive.

Between 1985 and 2000 potential buyers generally avoided the small units, and with good reason. During that period a few friends and strangers asked me if it would be wise to invest in a small wind turbine. My answer was that unless they had the ability to repair it and would not mind spending the weekend hanging from an 80-foot tower, they might want to pass on any opportunities. I was really saying: don't be a fool. This is not an easy technology, although it seems deceptively so. If one wants to make a difference, remember that it is much easier to save energy than to produce it. Read up on wind energy. What is the cost-benefit breakdown? How about getting along with the neighbors? What about zoning? What size and style should I consider? Does my site have the necessary wind regime? What about photovoltaics? There have been many advances with solar panels. Perhaps solar would be a better choice? Or better yet, what about a hybrid system, taking advantage of both sun and wind? Do I want to stay on the grid or get off? What are the pitfalls of putting up a turbine? There are many questions that a potential purchaser must ask. I advise settling down with a yellow highlighter and Paul Gipe's *Wind Power: Renewable Energy for Home, Farm and Business* for a few days. Supplement Gipe with Mick Sagrillo, a small wind power expert from Wisconsin who has written many articles for his website and AWEA publications. These two men are wind energy enthusiasts, but they do not sell any products. They are honest in their appraisals.

What Is Needed: Standards

A lot of junk turbines entered the market in the early 1980s when subsidies were easily secured. In Wisconsin, for instance, the federal government subsidized to the tune of 40 percent, while the state kicked in another 26 percent tax credit. With a 66 percent write-off, both turbine producers and enthusiastic consumers jumped into the game. Unfortunately, whereas Wisconsin legislators designed the tax credits to encourage reliable wind turbine growth, it did just the opposite. Quality was missing. Sagrillo notes that "of the 80 or so wind turbine manufacturers that set up shop and sold wind turbines back then, only a handful are still in existence and manufacturing wind systems. Most designs had not been field tested, had fatal flaws, and failed within a year." Michael Bergey, a manufacturer of quality small turbines, has written: "A wind turbine is a deceptively difficult product to develop and many of the early units were not very reliable."[14] When tax credits expired on December 31, 1985, only his company and one other survived to 1986. The small wind energy upturn came to "a crashing halt."[15]

Government support and subsidies are back, but that does not mean we can relax our guard against machines that come and go almost as quickly as yearly automobile models. Sagrillo is particularly wary of the Chinese turbines that have flooded the market. They are inexpensive and are promoted on the Internet and elsewhere with enthusiastic trade literature and plenty of endorsements, but often they simply do not work. Cheap in price, they are cheap in quality. Sagrillo strongly recommends that buyers have assurances of technical support close at hand. He also questions turbines that are touted as "urban turbines," designed to be installed on a homeowner's roof. A rooftop turbine can cause vibrations at best, serious structural problems at worst. Engineers design turbines to be placed on towers.

A Better Design?

The design of small turbines is a subject of endless speculation. I include myself among those who feel that engineers ought to be able to devise a small wind turbine that is less intrusive than the standard three-bladed unit. Such a unit would expand the market. The industry needs a design that does not offend the neighbors who will be looking at it. The Skystream is a fine unit, but its curved blades are displeasing to the eye as it claws at the sky. Engineers designed the blades to be efficient in drawing power from the wind, but I believe it sometimes makes sense to allow aesthetics to compromise efficiency, by up to perhaps 20 percent, and I believe purchasers should have a choice.

Inventors would love to saturate the market with small generators, satisfying that need for choice.[16] A few years ago one such inventor contacted me in Jackson Hole, Wyoming, and asked that I examine his prototype of a rooftop vertical axis unit he had set up in a home garage. I was encouraged that it was an enclosed unit with no visible movement. From an aesthetic viewpoint, it had potential. In the garage he and a partner turned on a fan that activated the turbine. It appeared to have some potential and he had patented his idea, but clearly it needed field testing by NREL or some reputable university engineering department. He asked for help. I offered to do what I could in exchange for 1 percent of his profits if he sold the idea. He said that could mean $100 million or more, a sum he would not consider; I decided against capping my fee at $100,000 and merely wished him well. He had lost contact with reality. He believed that soon every rooftop in America would be blessed (and vibrating) with one of his turbines. In truth, as Paul Gipe has often noted, when it comes to small turbines you have to look far and wide for a workable new idea.

Small vertical axis turbines are available, however. Sigurd Savonius, a Finn, invented the S-shaped vertical turbine in 1924, and Finland has been enamored with them ever since. They are artistic, silent, easy on birds, and practically indestructible. However, their power production is less than that of the horizontal style. Today the Windside Production Company offers eight vertical turbine models. In the United States, Helix Wind, a global energy firm founded in 2006 in San Diego, has built a similar vertical turbine billed as having a "seashell shape." It is now in limited production; recent passage of the tax credits for small turbines should help. Helix Wind claims a number of technical advantages, and Savonius-type units have the environmental benefits of being easier on the eye and more bird and bat friendly than traditional styles. Yet Helix freely admits their vertical turbines only produce 50 percent in power output compared to an equivalent bladed turbine.[17] A truly efficient, reliable system has eluded us, remaining a challenge for engineering schools and creative students.

Let the Buyer Beware

In this creative free-for-all, what really works? At the moment it is hard to tell. There are no established standards. Mick Sagrillo believes there should be. While large turbines are now well established as serious players in the energy marketplace, small wind turbines remain mostly a curiosity, and "unreliable, untested, or half-baked designs" have been foisted on an unaware public. Reporting on small turbine testing done in the United Kingdom, Sagrillo noted the tremendous disparity between manufacturers' promises and test results. Encraft, the testing consultant company, found that the small turbines produced only 25 to 50 percent of the expected output. All this exaggeration represented a failure of manufacturers' "to market their products ethically." Some cross the line to "outright fraud."[18] Sagrillo notes that the state of Massachusetts put a hold on its small wind incentive program because none of the turbines were performing as expected or as advertised.

Sagrillo is convinced that the industry should have product certification. The certifying organization would give accurate information regarding reliability, annual energy output, sound emissions, and safety mechanisms. The buyer would know what to expect, as in automobile milage standards. Certification would prevent what Sagrillo calls "rated capacity creep," the proclivity of a manufacturer to be overly optimistic regarding output. Such a deceptive policy not only helps to sell the turbine but also allows the purchaser to maximize tax credits.[19] For my own part I would only recommend Bergey Windpower and Southwest Windpower. Mike Bergey and his father, Carl, have been building windmills since the mid-1970s. They know turbines, and they offer a five-year limited warranty plus their own reputation of making it right with the customer. Southwest Windpower has been making small wind turbines since the mid-1980s. Andy Kruse, one of the Southwest founders, was young, talented, and committed when I met him at the AWEA 1987 annual conference in San Francisco. Southwest crafts a number of products such as their AirBreeze to provide a little power for sea-going sailboats, their Whisper series for small homes, and their state-of-the-art Skystream turbines, providing 2.4 kW output with a 115-foot swept area. Like the Bergey machines, it also has a five-year limited warranty. In the absence of certification standards, an established manufacturer offering follow-up capability is the safest bet for buyers.

One Woman's Pride

To look at a specific case, Mary Starrett of Prosper, in Collin County, Texas, takes great pride in her new wind turbine. In an effort to lower the utility bill on her 2,200- square-foot home she installed a Skystream 2.4 kWh on a 33-foot tower, costing $13,500. In May and June of 2008 her utility bill averaged $93, while a neighbor with a similar home laid out $350 in May. Whether the turbine will pay for itself depends on her utility rates and the turbine's reliability. Repairs can eat up any savings in a hurry. In the meantime, however, Starrett and her daughters have a new attitude toward the wind: "We watch [the turbine] all the time. It is better than watching TV." They are most pleased when the wind is blowing, and the turbine is turning. Not surprisingly, she has had many inquiries about her new backyard addition, and Charles Crumplet, her installer, reports that he is installing an average of one turbine per day.

The Starrett turbine has also triggered another issue. When are town turbines appropriate and when are they not? The town of Prosper approved Starrett's turbine as an "accessory structure," but the possibility of more turbines prompted the town council to debate what should be allowed, a debate going on in towns across the country. In Prosper the new code limits a turbine to 60 feet in height, including the blade. It must be placed at least 75 feet from any adjacent property line. It cannot be placed in the front yard, and the lot must be at least an acre in size.[20]

It is obvious that wind turbines are not appropriate for town subdivisions, but in more rural situations, a turbine can be an environmental benefit as well as an economic investment. At present there are 30 percent federal tax credits available, although Mary Starrett will not receive all of that advantage. Yet she enjoys the psychological benefit of seeing the turbine at work to lower her utility bill. The machine provides meaningful satisfaction; seeing that spinning turbine gives her a boost.

In Flagstaff, Arizona, the City Council has wrestled over an ordinance regulating small turbines (as tall as 100 feet). Recently they decided to grant permits on lots as small as half an acre, although not in residential zones. But how about residents who live adjacent to commercial areas? Councilman Jim McCarthy opposed the ordinance, noting that "there are many areas of the city that are, in fact, residential [but are] zoned commercial." Perhaps taking a more practical approach, he stated that the city "should not allow people to do things that are completely inappropriate, not just rely on zoning." We can expand on Councilman McCarthy's admonition. A wind turbine in a subdivision can divide a community. I know a couple who insisted on erecting a turbine in spite of the neighbor's objections. The neighbor did not go to court but rather spent about $2,000 on landscaping to protect his view. It is not worth such animosity. Doing everything possible to conserve energy use in one's home is better than to producing a little electricity at the expense of neighborhood harmony.

Your Own Wind Turbine

If you have any idea of owning your own wind turbine like Mary Starrett, do your homework. Making the wrong move in this tricky business can bring on much grief. As mentioned, nearby residents should be consulted. Without their consent one risks at least dirty looks and the silent treatment. The proper permits must be in order. Choosing a turbine requires rejecting poorly constructed machines and dealers who disappear after you have put down your money—you want a reliable dealer who can service the turbine when needed. As Paul Gipe says, "You don't want a dealer who lives on the other side of the continent."[21] A wind turbine must run year after year if the investment is to pay off. It would be enlightening to find out how Mary Starrett feels after her turbine has been spinning for several years.

Some Big Houses Stand Alone

When we think of a bold person getting off the grid with a stand-alone wind system, we have a vision of isolated mountains or canyons and tiny cabins featuring 50-degree temperatures: we imagine that people who are off the grid are roughing it because central power is not available. But we should not assume that wind energy is only for individuals who enjoy cutting wood for warmth and who live with just a bare minimum of light-bulbs. Wealthier patrons are building "net zero" homes that are neither small nor modest but are energy efficient. These homes use energy-saving construction and hybrid renewable energy systems, often a combination of solar panels and a wind turbine. When Brett Nave and Lori Ryker built a 2,900-square-foot residence north of Livingston, Montana, they decided not to pay Montana Power the $20,000 necessary to run a power line to the site. A 2-kilowatt turbine captures the notoriously strong Livingston winds, and twelve solar panels utilize the abundant sun. They rarely use the backup generator. There is satisfaction in constructing a sustainable dwelling that uses only renewable energy to meet the needs of its owners comfortably; the drawback is that only the affluent can usually afford to build such a home.[22]

In rural Almonte, in the province of Ontario, William Kemp lives very comfortably off the grid. He decided that $13,000 (Canadian) was more than he wanted to pay for a hookup, so he worked out how to keep his 3,000-square-foot home warm and have plenty of electric appliances providing amenities. He uses a turbine perched on a 100-foot tower and a 1,200-watt photovoltaic panel. For heat he uses a catalytic wood stove and an efficient propane fireplace.

Kemp is adamant that an energy-efficient home is the key. In the tough Canadian climate, he says energy consumption can increase by five to ten times in a leaky home. He demands a well-sealed house with the most efficient appliances, and he and his wife are happy with electricity from their Bergey wind turbine, photovoltaic panels, and their sizeable battery bank to provide needed storage. One might think they live a Spartan existence, but unless he is fooling us, they have all the electrical gadgets they want.[23] Living comfortably off grid takes planning and effort, but for increasing numbers of people the satisfactions of helping the planet and living independently are worth it.

Doing Good for People as Well as Earth

Millions of people in underdeveloped countries have no electricity. Obviously they are surviving and their happiness is not dependent on electricity or material goods. Yet few would deny that an electric light to read by can open new horizons. As a historian I often wonder how nineteenth-century Americans read for hours by the light of candles or kerosene lanterns. By all accounts, when people received electric lights, their lives changed for the better.

Water-pumping windmills have been one of America's most successful exports. For one hundred years, companies have shipped their windmills throughout the world. They are still valued structures in many villages because availability of clean ground water eases life greatly. Now the new kind of windmill called the wind turbine has been making an appearance in developing countries. Some countries, such as India, have installed large commercial turbines, but the power goes to the wealthy cities, leaving rural communities unchanged.

Small, isolated communities cannot afford the tremendous cost of a commercial-sized wind turbine. Usually a 2 MW turbine costs $2 million to $3 million. Besides the initial cost, the expense of maintenance would break the budget of a struggling town. A large turbine is not the answer. However, a small turbine with a few batteries can power lighting, radios, and a few television sets. Some might argue that unenlightened people are better off without electricity, particularly television, but no one can deny the importance of power for communication, medical assistance, and education. For these people a small wind power unit can be the answer. If it is combined with solar panels, the village can have electricity much of the time. Adding a backup diesel or gasoline generator and batteries can produce a reasonably reliable 24/7 electrical system.

Relying on a diesel or gasoline generator, however, defeats much of the purpose of a wind and solar hybrid system; refined fuel requires cash, robbing poor communities of self-sufficiency. Nonprofit agencies have installed diesel generators, but in the long run the costs of fuel and maintenance have undermined their use. Wind turbines and solar panels do require maintenance, but the fuel is free, making wind turbines an appropriate technology in some remote regions.

Reaching Out

An American company doing significant reaching out is Bergey Windpower. For years the company has exported its small generators to remote places of the globe. The company has more than fifty representatives on five continents. Bergey makes a special effort to see that their turbines are trouble free for at least five years. Their 1.5 kW BWC 1500 is designed for what the company calls "village level operation and maintenance," following guidelines established by the World Bank.

Some of the company's completed projects give an idea of the work it is doing. Jengging village in India is deep in the foothills of the Himalayan Mountains. Since it is doubtful that utility lines would ever reach Jengging, in 1988 the Indian government erected two Bergey 10 kW units. They still provide the village with electricity. In Morocco, with support from USAID, the government installed wind-electric pumping system to replace diesel-powered pumps; the diesel fuel was simply too expensive. Also in Morocco, two 10 kW turbines supply electricity to provide water for four villages in the Naima Rural Commune. The turbines replaced diesel units, which again the villagers could not sustain. To look at the economy of scale in small turbines, consider China. In Inner Mongolia some one hundred thousand locally manufactured small turbines provide a capacity of 10 MW of energy, usually powering the limited electrical needs of Chinese and Mongolian tribal yurts. This is the largest decentralized electrification project in the world.[24]

One hundred years ago American companies exported thousands of water-pumping windmills to undeveloped countries. Now wind turbines are migrating across the globe. These Africans in Kenya are anticipating a little electricity to charge their cell phones. (Bergey Windpower photo.)

Small wind turbines can bring a little electricity to remote villages. This Bergey Excel 10 kW turbine is one of ten raised near Panjshir, Afghanistan, to improve villagers' lives. The U.S.-funded project is a small price to pay in a land where our goodwill is in short supply. (Bergey Windpower photo.)

Perhaps the most interesting international wind project is in Afghanistan. In a mountainous country long raked by the ravages of war, a central grid system is problematic. The U.S. Army contracted with Bergey to install ten Excel R turbines in the mountainous region of northern Afghanistan near the city of Panjshir. Bergey's windpower dealer in Kabul planned and constructed the project, and it was dedicated in November 2008.[25] It is funded by American taxpayer dollars on the basis that a small investment will have big returns in terms of winning the minds and hearts of the Afghan people.

For a time in our history, small wind turbines seemed to be going the way of the dinosaur. From 1956 to 1975 no American company manufactured a small wind turbine. The resurrection came in the 1980s, when increasing numbers of people realized the economy of living off the grid as well as the benefits of supplementing the grid power system. New companies came and went. The technology still is not simple to master, and yet, it is an appropriate technology, reflecting the environmental concerns of the world in which we live. Increasingly, the small turbine has found a niche in the United States, and certainly in the emerging nations of the world, where the benefits of a free energy source are fully appreciated.

CHAPTER 9

Other Solutions

Without fossil fuels, we'd return to the stone age.

Windmills and warm sweaters will save the planet.

— Vijay Vaitheeswaran on popular myths, *Power to the People*

Experts believe that wind energy will never exceed 25 percent of U.S. power needs. As noted, some regions, particularly in the South, lack the wind resource necessary for efficient production. Some places that do have the requisite wind may have residents who object to having specific landscapes packed with wind turbines. But the primary problem everywhere is that the wind does not blow all the time. Unless Americans are willing to live without guaranteed electricity on demand, a system relying solely on wind cannot exist. Inventors such as Charles Brush and thousands of isolated farmers and ranchers who used wind turbines depended on a bank of batteries to carry them over during the doldrums. However, filling the basements of American homes with storage batteries is problematic at best.

For all its advantages, wind energy seems destined always to supplement power sources that engineers and utility companies can control as the base load. In this chapter I review some of the options available to utility companies, both private and public, and explore the various ways in which we generate electricity. Many books have been written on each of these energy sources. Wind energy is just part of a selection of generation options. Each source makes a contribution and contributes to the diversity of our power sources. Diversity is an important concept, as are stability and balance. From an environmental perspective, however, not all generation methods are equal. We all need to contemplate which energy production systems should be favored, particularly by government programs. In the next decade, as the United States strives for energy independence, we will make public policy decisions to reflect both our needs and our commitment to stewardship of the land and the earth.

Hydro Power

Hydro power is one of the oldest sources of mechanical energy humans have developed.[1] In Europe the waterwheel can be traced to the inventiveness of ancient Greeks; by the time of Christ water power was in common use. For centuries European millers and manufacturers used improved waterwheels to grind grain and perform many industrial tasks. The waterwheel was particularly well adapted to England and northern European countries, where rivers and streams flowed year round, offering abundant opportunity. Waterwheels migrated to the American colonies. One expert notes that the 7,500 waterwheels extant in 1790 grew to 55,000 by 1840. By 1850 advances in steam power began to push the waterwheels aside. Perhaps the most elaborate industrial use of water power in the United States occurred at Lowell, Massachusetts, where a complex system of hydro power provided rotary energy to power eighty-eight looms, which in turn fashioned textiles for the nation. Historically, most waterwheels were either "overshot" or "undershot," depending on whether the falling water dropped into buckets or pushed spinning paddles from below.[2]

The principle was similar to that of wind power in that both power sources occur in nature and are renewable. Wind power employs air movement caused by changing atmospheric pressure, while water power utilizes gravity and falling water (weight, overshot) or the impact of rapidly moving water (undershot). Both wind and water turned a rotary shaft, which was then used to turn mechanical equipment. The great advantage of water over wind is that it is concentrated rather than diffuse, thus accomplishing more work. Because water is concentrated and confined it was also more manageable for human uses, often creating energy monopolies based on riparian water rights.

Another advantage is that water can represent stored energy. Grain millers and manufacturers found that by establishing a crude dam upstream from the waterwheel they could release the water when it was needed. By controlling the storage and release a miller might obtain optimum use of the energy source. The head gates could be closed at night, and water could be released when the mill was ready to work in the morning. This principle likewise drives hydro power generation.

Few waterwheels survived the industrial revolution. Those that turn today are mostly novelties for the tourist trade: symbols of an earlier age. In our modern age, water power is associated with dams, some of them huge. These dams and their blockhouses of hydro turbines account for approximately 10 to 12 percent of the nation's power.[3] Besides generating power, dams store water for irrigation, provide protection from floods, and provide boating and recreational opportunities for an affluent public.

The evolution of large dams is associated with the mining industry in California. Water proved essential to all aspects of mining. Once argonauts had panned the easily gathered gold, it was time to wash down the gold-bearing mountains. This was accomplished with high-pressure hoses, capturing as much of the kinetic energy of water as possible. Such a use of water power proved an environmental disaster, sending debris, gravel, and soil downstream to where it was not wanted. The practice was finally outlawed in California in 1887.[4]

Successful hydraulic mining required strong water pressure. This water pressure was created by storage dams and then pipes (penstocks) that would drop at least 1,000 feet of gradient, creating as much pressure as a ton per square inch. Water leaving the nozzle could exceed 200 miles per hour, making today's fire hose seem like a child's toy. At this pressure even six or eight workers could not hold the nozzle. It had to be anchored by steel and iron bolts.[5]

In essence, this technology to wash away mountains is closely related to the modern hydroelectric turbines. Water stored behind dams is released into a pipe (penstock) that drops significantly in elevation. Under extreme pressure, water jets hit the cups or blades of a Pelton or Francis turbine, turning it at up to 500 revolutions per minute. One expert states that a well-designed turbine "can convert more than 90% of the kinetic energy of the water into rotary motion, at high rates of revolution." This motion converts the energy to electricity. It is extremely efficient.

With water power supplying some 12 percent of United States electrical energy and about 20 percent of world energy, why not simply increase this figure? The answer, of course, is multicausal. Most of the best sites have been developed with colossal dams, many in the American West. The sites that remain often do not meet reasonable cost-benefit ratios. Ironically, while hydro power is environmentally benign (producing no heat or CO_2), environmentalists fiercely defend the remaining free-flowing rivers. These river defenders find allies among people whose homes, livelihood, or recreational activities would be displaced if the river became a reservoir. There are few good sites left, and those that might seem available are fiercely defended by interest groups. Even the Bureau of Reclamation, the traditional dam-building arm of the government, acknowledged in 1987 that its days of dam building were over, and it would focus on water conservation and water quality issues. Thus, while wind energy is on a great growth spurt, hydro power is static and aging. It is possible that in this century more dams will be dismantled than built. At least in the United States, we cannot expect our energy problems to be solved through water power. Significant hydro power projects are under way in other parts of the world, but it is unlikely that engineers will design and build new high dams in the United States.

On a small scale, hydro power can provide electrical storage. The nation has more than thirty small low/high reservoir systems, generally called pumped storage hydro systems. At night when electricity is cheap the operators activate the pumps to move water from the low to the high reservoir. Usually in the afternoon the operators release the water into a penstock, where it rushes down to a turbine and generator system, creating electricity at a high value per kilowatt during the time of peak demand. Reservoirs, large and small, are one way to store electricity until it is needed. In some regions hydro power can be combined with wind turbines to create a viable renewable system.

Coal

Coal is the workhorse of electrical generation. In the twentieth century burning this abundant resource provided heat that produced steam, which turned the generators. In the twenty-first century coal is still the natural resource that provides over 48 percent of the electricity we use. Coal fuels most of the large base load generation facilities in the United States. It is relatively cheap and, most important, it is plentiful. In effect, coal is stored solar energy, the result of millions of years of accumulated plant and animal matter crushed solid by the weight of the earth. The advantage of a coal fuel plant is reliability. Utility engineers can ramp up a 1,000 MW plant to its full capacity and run it indefinitely, whereas wind energy is at the mercy of the elements, regardless of electrical needs or peak demand periods. Hence, in spite of its clear disadvantages, we are beholden to dirty old coal for the electricity we want.

The Union of Concerned Scientists made a revealing comparison of coal power versus wind power. Coal generation causes smog, soot, and acid rain as well as heat (causing global warming) and toxic chemicals. Getting rid of the ash and sludge create more environmental problems. Mining, transporting, and storing coal uses massive amounts of energy, while polluting land and water. Furthermore, the actual coal electrical plant uses massive amounts of water (2.2 million gallons per year for a 500 MW coal plant), and the water returned to its source is 20 to 25 degrees warmer. In contrast to these dismal realities, wind energy was the clear winner. The Union of Concerned Scientists reiterated what we already know. Wind energy gives off no air emissions. It uses no fuel that must be mined or transported. It does not heat or pollute water. In fact, it uses no water and it creates no wastes. It does create some pollution and use some natural resources when wind turbines are manufactured, but no more than other energy options do.[6]

The pollution figures for the approximately six hundred coal plants in the United States are startling. According to one source, coal plants are responsible for 93 percent of the sulfur dioxide and 80 percent of the nitrogen oxide emissions spawned by the total electric utility industry. Coal plants contribute significantly to global warming and emit 73 percent of the carbon dioxide spewed into the atmosphere by electric generators. Much of this pollution is from older plants, where owners have been able to avoid Clean Air Act standards through variances. However, sometimes even variances cannot save a plant. On January 1, 2006, the Southern California Edison Company closed down its Mojave Generating Station in Laughlin, Nevada, rather than pay $1 billion for environmental upgrades. But even though the percentage of coal-generated electricity is slightly down (48.5 percent to 48.2 percent), the closing of the Mojave Generating plant is unique. We still need coal plants, old and new.

Coal will be with us for decades, and those in the business are doing what they can to make it more environmentally friendly. Research to produce "clean coal" is under way, and it seems that most of the sulfur dioxide and nitrogen oxides can be removed before burning. Carbon dioxide presents a more difficult technological challenge. Obviously, coal power stations introduce hazardous byproducts into our environment, and the concept of clean coal may only be a concept. There is plenty of evidence to support the warning of Dan Becker, director of the Sierra Club's Global Warning and Energy Program, in his belief that "there is no such thing as 'clean coal' and there never will be. It's an oxymoron."[7]

Even if scientists perfect clean coal technology, the question of cost remains. The advantage of coal has always been its availability and its low cost. Such a costly procedure as carbon sequestration (depositing the CO_2 underground) to burn clean coal may price this energy source right out of the market. In the near future, coal's competitor will be nuclear power, and perhaps natural gas. The cost of nuclear fuel will surely rise dramatically, but whether a new environmentally acceptable coal will be competitive with yellow cake and fuel rods remains to be seen. Either way, the price of electricity will be going up.

With such a universal usage and such long and profitable track record, one would think that federal subsidies would not be needed for the coal industry, and yet they are. Research into clean coal is largely subsidized by the Department of Energy. Secretary of Energy Steven Chu acknowledges the abundance and the importance of coal. He realizes it is a dirty fuel, but he is committed to cleaning it up. "It is imperative," says Chu, "that we figure out a way to use coal as cleanly as possible." It is a necessary evil, and thus far research to cut sulfur, nitrogen, and mercury pollutants has had only limited success. Carbon dioxide continues to be a stumbling block, but with an additional $800 million added to the pot through the 2009 stimulus bill, perhaps the utility companies can make progress.[8] We cannot clean up the air without it.

Additional coal industry subsidies generally escape our notice. Perhaps the most costly has been the federal government payout to coal miners who suffer from black lung disease. Over the last thirty years payments have been in the billions of dollars to cover medical costs.[9] Obviously there is justice and necessity in this subsidy, and yet it is not reflected in the true cost of coal. We must remind ourselves that wind, contrary to coal, comes to the generating plant without environmental or health issues.

Oil

No one would question the importance of oil to the energy production of the United States. However, when it comes to petroleum as a provider of electricity, it pales beside other sources. It was not always that way. In

1973 fuel oil produced 16.8 percent of the nation's electricity. By 1983 that figure had dropped to 6.2 percent, and by 2006 only 1.6 percent of electricity came from power plants burning petroleum.[10] That figure is not likely to go up.

The reasons for the rapid decline of oil are easily understandable. As natural gas prices decreased, petroleum was on an upward trajectory, being priced out of the market. Economic issues combined with emission restrictions to change the playing field, forcing many plants to switch from oil to natural gas. From 1980 to the turn of the century, plants burning the heavy fuel called number 6 fuel oil became a symbol of dirty oil, since the plants emitted dark smoke, a clear contrast with natural gas. Furthermore, the oil has a high sulfur content, not only polluting the air but having a corrosive effect on heating systems, shortening their life spans. Many of these old oil-burning plants have been closed and demolished. It is unlikely there will be any comeback for oil.

Natural Gas

Natural gas has largely replaced oil and has become one of the workhorses of electrical generation, along with coal. Two-cycle gas turbines are most often used in conjunction with a base load coal plant. When demand increases, the natural gas turbines kick in to meet the megawatt increase.

Depending on what statistics are consulted, natural gas produces something over 21 percent of the nation's electricity. It is abundant, relatively clean, and affordable, but it does produce heat. As anyone who uses gas appliances at home knows, natural gas prices vary rather dramatically from year to year. However, companies have developed a number of new gas fields in the last few years, meeting demand and lessening the chance of a dramatic price spike like that of 2008. Although natural gas and wind energy are both peaking sources (that is, supplementing a base load plant), the potential for natural gas is great, and new wells are coming on line throughout the country.

Natural gas is used not only in households and turbines but also in transportation. Many cities have switched their bus fleets and other vehicles from gasoline to natural gas. This trend will continue and, ironically, could greatly increase the use of wind power. The plan of Texas oil billionaire T. Boone Pickens to rid the nation of its dependence on foreign oil involves replacing the 20 percent or more of power generation from natural gas with thousands of wind turbines placed in the windy corridor known as the Great Plains. If wind energy replaces natural gas, what do we do with the natural gas? The Pickens plan calls for converting to natural gas all the diesel-powered eighteen-wheelers that criss-cross the country, since 20 percent of every imported barrel of oil goes to fuel those trucks. The plan starts with trucks and eventually expands to automobiles.[11]

The plan envisions that electrical generation will no longer use natural gas, but is this necessary? If we are to believe the natural gas industry, there is at least a fifty-year supply, and new fields seem to be continually opening. It might seem more logical, and environmentally sound, to reduce the use of coal and continue the 20 percent use of natural gas.

Solar Power

Although wind power is simply a variation of solar power, let us briefly look at energy *directly produced* by the sun. Probably all of us have at some stage captured the sun's energy with a magnifying glass focused on paper to create a fire, or at least scorching whisps of smoke. Solar power today is a sophisticated matter of flat panel solar collectors on home and store roofs, a sight that was rare twenty-five years ago. Great progress has been made in the cost and efficiency of these collectors. Photovoltaic solar cells are now becoming competitive with grid-delivered electricity in rural areas, and they are much better suited for use in urban areas than are wind turbines. Few neighbors would tolerate a 70-foot wind turbine tower next door, but a half dozen rooftop solar panels normally do not pose a problem; I have a nearby neighbor who combines the two, using a 1 kW Bergey wind turbine and six solar panels for much of his power.

Besides small flat panels, there are a few much larger solar power systems that use lenses or mirrors to focus a large expanse of sunlight into a small beam. Some years ago as I was approaching Tehachapi Pass from the Mojave Desert, I drove my car over a small rise and was met by a blinding light that forced me to pull over. It was a Solar Energy Generating Systems plant, which is still functioning as a heat source to generate electricity.

Solar energy is so universal and has been so broadly used historically that we cannot but predict expanded use in the future. Much of the application is low tech, and in an energy-aware society it is just common sense, as in the Anasazi inhabitants of the Southwest favoring a south-facing cave to survive a cold winter. Today, in most regions of the country a passive solar house is a logical design that can be built with little added construction cost.

Thus employing the sun in sensible ways is much simpler then erecting wind turbines. Both solar and wind power have that wonderful attribute of using inexhaustible natural energy sources. Like wind, sunshine does not pollute, and it costs nothing to get the solar fuel to the generating plant. At the moment solar is widely used for individual water heating and for providing small amounts of power in off-grid situations. However, it is commanding more and more of the market for on-grid installation, although the cost of a kilowatt-hour of solar power is still higher than for wind power.[12] Wind energy is better suited than solar energy for large-scale electricity production that employs the grid for distribution.

Nuclear Energy

Had the prognosticators after World War II been correct, we would now all be happily (or fearfully) powering our homes and factories on nuclear energy: electricity they claimed would be so cheap and abundant that we would hardly have to meter it. Utility companies hailed President Dwight Eisenhower's 1951 "Atoms for Peace" initiative as a new age. There would certainly be no need for wind energy. That was the decade in which wind energy research came to a halt in the United States and every small wind energy company closed its doors. For more than a decade, nuclear power seemed to hold the answer, and the Atomic Energy Commission invested some $27 billion in the development of nuclear power between 1955 and 1964. During the 1950s private capital contributed only 12 percent toward government-sponsored nuclear reactor programs. Commercial nuclear power received a subsidy unparalleled in the American experience.[13]

With such government assistance, private utilities ordered some 231 nuclear plants, to be delivered by 1974; those were halcyon days for nuclear power. But plans did not produce plants. Many orders were canceled and utility owners even abandoned partially completed units. Cost overruns were phenomenal. The end came in 1983 when the Washington Public Power Supply System (WPPSS), which planned to build five nuclear plants, declared bankruptcy and defaulted on $2.25 billion dollars' worth of bonds. One critic of nuclear power remarked that WPPSS "promised power without cost, and they delivered cost without power."[14]

Public fear was a key factor. The Three Mile Island incident in 1979—in which many of the vaunted safety features of a nuclear plant failed, causing radiation leakage and a near meltdown—shattered public confidence. A few years later, when the Chernobyl nuclear facility in the

Ukraine exploded and spread deadly radiation throughout a vast area, the days of nuclear power seemed over.

Yet in spite of both failure and fear, nuclear power is responsible for almost 20 percent of the nation's electrical generating capacity. The percentage is not comparable to that in Japan or in France, which today relies on nuclear facilities for over 70 percent of its electricity. France has had no significant accidents in more than fifty years of nuclear generation. With such an example, the idea of nuclear power plants has been resurrected and the federal government is once again offering significant subsidies. Combining the fact that a nuclear plant's routine operation emits no pollutants into the air and the dispersal of produced heat through large bodies of water makes nuclear energy appealing in a world concerned with global warming. If utility companies do build new nuclear plants, and they probably will, these will be costly. And if wind energy is considered undependable, nuclear energy is not much better. According to one source, of all the 132 nuclear plants that have been built in the United States, 21 percent were permanently closed due to unreliability or cost problems, while another 27 percent have completely failed for a year or more at least once.[15]

However, the technology has advanced. Uranium fuel can now be used more efficiently than was once the case. Yet neither the French nor any other nation has resolved the fearful radioactive waste problem. No matter where the federal government finally decides to store spent rods, managing the waste will be both controversial and expensive. It is one of the externalities that the taxpayer must bear, and that makes nuclear power expensive, and that continues to provoke a widespread adverse attitude toward nuclear power. Many people are counting on new technology with hydrogen and fuel cells to provide the electricity of the future.

As far as wind energy is concerned, the two power sources are not in conflict or competition. Nuclear power is suitable for base load power plant operations. If new plants are built, they will likely replace coal operations.

Geothermal Energy

Few sources of power are as environmentally friendly as wind, but geothermal energy is one of them. Its energy source is the heat of the earth. It emits almost no greenhouse gases, just water and steam. According to the Geothermal Energy Association, geothermal production is the fourth highest source of renewable energy in the United States. It provides power for 2.4 million homes and has 3,000 MW capacity.[16] Almost all this activity is in California, and specifically at The Geysers in the northern part of the state. Although wind and solar power receive most of the publicity, in California geothermal plants produce some 5 percent of the state's energy, more than solar and wind combined.

Geothermal power cannot be sidetracked by weather. Unlike solar and wind energy, the heat of the earth can produce power every day, all day. With such attractive features, why has geothermal power not had a greater impact? The difficulty rests with geology. Much of the country simply does not have an accessible resource. Some of the areas that do, such as the huge Yellowstone caldera, are off limits to development. We need not fear energy producers disturbing the complex underground network that supplies heated water to Old Faithful geyser and the thermal features, drawing millions of visitors yearly. In the western states plus Alaska and Hawaii engineers are developing almost 4,000 MW of new geothermal power. This development has been aided by the Geothermal Steam Act Amendments, passed by Congress in 2005, providing incentives similar to those for other renewable energy resources.[17]

An aspect of geothermal energy is employed in many thousands of American homes via a heat pump, utilizing the natural heat of the earth to heat or cool a building. The heat pump exploits the temperature difference between the earth's surface and the air. For instance, in winter in Wyoming, surface air temperature may be 10 degrees below zero, while the temperature 4 feet beneath the ground is 55 degrees. Conversely, summer air at Carlsbad Caverns can be over 100 degrees, while deep in the cavern the temperature is a pleasant 72 degrees. Exploiting such differences in temperature for the comfort of people in a building is the job of the heat pump. Heat pumps are, of course, environmentally beneficial. They use no fuel and only a little electricity. Because they are so environmentally friendly, the Environmental Protection Agency considers heat pumps an excellent home investment, and the American Recovery and Reinvestment Act of 2009 provides a tax credit of up to a 30 percent of their cost. With these clear advantages, the heat pump industry has been growing at approximately 15 percent each year, and there are now some eight hundred thousand pumps at work across the country.[18]

Clearly geothermal energy is an important player. In production of power, installations in California provide for the electrical needs of 2.4 million homes. The figure will grow. Heat pumps are a different category.

The Geysers geothermal plant near Santa Rosa, California, is the largest geothermal plant in the world. (USGS–NREL photo.)

They do not produce electricity as wind turbines do, but they do conserve electricity by taking advantage of our earth's natural processes.

Fuel Cells

One of energy's hot topics is fuels cells. Some scientists view as the answer to our energy needs, although they are not an energy source. Vijay Vaitheeswaran, environmental author and journalist for *The Economist*, provides an abbreviated definition: "So what are fuel cells? In a nutshell, they are big batteries that run for as long as fuel is supplied. There are various types, but nearly all work by combining hydrogen with oxygen to produce electricity, while resultant emissions are no worse than water and heat."[19] The "big battery" does require an energy source. The electricity then produces a current that can power a home, a computer, or an automobile.

The concept of fuel cells goes back 150 years, but no scientist could translate theory into a practical product. The American space program changed that. With NASA's need for a lightweight, reliable power source, fuel cells did the job. However, they were expensive, and in the 1970s and 1980s they were in a price range that only NASA could afford. The next push came from California, which not only encouraged wind energy development but initiated a zero emission vehicle mandate, decreeing that by 2004 one-tenth of the cars on California highways should be zero emission cars. That did not happen, and in spite of the efforts of such brilliant environmental scientists as Amory Lovins, founder of the Rocky Mountain Institute, the "hypercar," as Lovins calls it, is not on the roadways.

Our primary interest is in what fuel cells can do for stationary power needs, mainly homes. Such companies as General Electric, Siemens, and United Technologies are deep into fuel cell research. It is possible that one day ours homes will be powered by a fuel cells the size of a dishwasher.[20] It would require an energy source, but it would be a stand-alone electrical system, totally divorced from the grid.

What goes around comes around, it seems. Advocates of central power destroyed the market for decentralized wind power. Now fuel cell technology reestablishes the viability of individualism and stand-alone power systems. There will likely be a place for wind energy if fuel cells represent our future energy supply. However, they are not here yet, neither for cars nor for homes. In the meantime, wind turbines continue to expand their electrical market share. Perhaps in time the old technology of the windmill will meet the futuristic technology of fuel cells, combining for a cleaner, sustainable, world.

Conservation

We would be remiss not to address the subject of conservation of electricity, the most important of alternatives. It is evident that production is determined by consumption. The greater the nation's demand, the more we must expand our means of production. Americans have a problem with consumption. Our appetite for electricity seems to know no end, mainly because we are so affluent. Two houses are a common luxury; at least one must have a hot tub; and we must be able to travel conveniently from one to the other, requiring two cars, or perhaps three. In the process, we use mountains of electricity. Even when we do not inhabit that second home, it consumes electricity. In Jackson Hole, where my wife, Sherry, and I have a cabin, some of the wealthy heat 8,000-square-foot homes even while they are wintering in the Bahamas. In a Wyoming winter you run up a weighty electrical bill just to keep a home at 50 degrees. The bill does not seem to matter much, nor does the consumption of electricity;

unaware of where the juice comes from, people look at me quizzically if I ask questions. If they can afford the power, what does it matter? Changing such behavior seems the very least we should be doing. While we stress production of electricity, consumption is the key. We must become more aware, for in the words of Secretary of Energy Steven Chu, conservation of energy is the "lowest hanging fruit." Energy production is difficult, while energy conservation is much easier.

In defense of the wealthy, some do realize where their power comes from, and they are beginning to pick and choose not only on the basis of economics but on the basis of the environment as well. In Jackson Hole one can purchase blocks of wind power from Lower Valley Power, our utility company. Taking this action is a "feel good" step, creating little real change in consumption. Nevertheless, if enough people participate, even this kind of conservation will have an effect on production. Creative conservation thinking saved California from a number of nuclear power plants in the late 1970s. California utility companies found they could make money by *not* producing electricity. Here was one of the first examples of what Amory Lovins designated as "the soft energy path," essentially a path to profit through environmental stewardship.[21]

Lovins and his Rocky Mountain Institute realize that most of our environmental choices are determined by economic factors. But again, that is changing. Take recycling, for instance. Across the United States families and individuals are recycling their waste even though there is little individual economic gain in recycling all the products and packaging of our throwaway culture. It just feels good, and it feels communal. If that is not the "correct" motivation, never mind.

Yet our per capita consumption of electricity has not diminished. On the contrary, it has increased. Electrical appliances and gadgets have proliferated into the few dozens. If you are old enough, just think about what you used in 1970 compared with today. The number of hot tubs in your acquaintance might be a good yardstick. Electricity makes our lives more convenient and more interesting, and whatever makes our lives meaningful is a worthy goal, but as far as electricity is concerned, there is a point of diminishing returns. If we have not already reached it, it may be coming.

In the meantime, how do we reverse the tide of electrical use? We need to do so if we want to spare the American landscape from being covered with wind turbines and nuclear reactors. It will not be easy. Conservation has been and will be important, and our appliances and gadgets are much more efficient than they used to be. But conservation alone will not do the job. We have to supplement conservation with economics. We must employ such basic concepts as supply and demand and the effects of scarcity. In a nutshell, the cost per kilowatt-hour of electricity must be much higher if we are ever to turn the curve downward. Here are a few suggestions:

Revisit the rate structure. In our capitalistic system the more we buy of any item, the less the cost. Buy one widget and the cost is $1; buy one hundred widgets and the cost is 50 cents per widget. The same principle has operated with utility companies. Rate structures have been geared to increase consumption by means of a declining scale that rewards high-volume use. By discounting the price at high levels of use, the system encourages waste and discourages conservation. In terms of social class, it compensates the wealthy, who may be purchasing electricity at say 5 cents per kWh, while penalizing the poor, who may be paying 10 cents or more per kWh.

Progressive utility companies recognize that a sliding scale encourages waste and penalizes the thrifty energy user. They have abandoned the declining scale for a uniform rate. No matter how many kilowatt-hours a household uses, the rate is the same. This system encourages wise use of the resource, while also promoting fairness. However, from an environmental viewpoint we must go further. Utility companies must determine what constitutes optimal or normal use of electricity for a household, and then place a surcharge on any electricity drawn beyond that limit. In effect, such a system would reverse the old norm. Now, the sliding scale is turned upside down, with elevated rates for foolish or unnecessary use. This is not as radical as it may appear. Water utilities regularly use three block rates: constant, declining, and increasing. They are self-defining and are based on average use of water. Those who use extra blocks of water will pay an increasing price. We need to recognize that electricity is both scarce and valuable.[22] The absent Jackson Hole homeowner may nevertheless heat a vacant home in the winter, but it will be very costly. That is as it should be.

I am quick to realize that such a rate scale presents problems and injustices, especially when it comes to businesses. But I believe that with effort, changing the rate structure can be made equitable. The gain for our environment will be worth the effort.

Install smart meters. The idea of smart meters has now reached the public. In Italy, much of California, portions of Texas, and numerous other utility regions across the country these clever meters now tell consumers how much power they are using and at what cost. The cost of this advanced metering system will raise rates at least 1 percent, but the savings will eventually lower costs for consumers. No longer will meter readers be necessary. More important, there will be at least a 25 percent savings for those customers who cut consumption during the critical peak periods of the day. PG&E, for instance, will set rates by season and time of day. The data are recorded by the utility company, but consumers can also view their energy consumption at home.

For those who wonder why we might want to have one, the primary reason is to save money. Your utility company must buy or acquire electricity at the "spot rate," which is largely determined by demand. At 4:00 P.M. in summer the rate is high because air conditioners will create a tremendous demand. By contrast, at 4:00 A.M. demand is low on the grid system. Electricity is not worth much at that quiet hour. Your smart meter will reflect this supply and demand curve and charge you accordingly. During peak demand hours, say 2:00 to 6:00 P.M., the meter will charge your account 12 cents per kWh. From 2:00 to 6:00 A.M. the meter will register just 6 cents per kWh. Will customers adjust their use to take advantage of lower rates? I believe they will, simply because consumers who adjust their use of electricity will be rewarded. Would such a smart meter change the nation's electrical habits? I believe it would. Why not run your dish washer in the middle of the night? Why not heat water or your house in the middle of the night? As earlier mentioned, plug-in electrical cars could become widely available, changing the patterns of our electrical use. I know one knowledgeable environmentalist who is eager to buy a plug-in car. He plans to charge it at night at 6 cents per kWh, drive two miles to work, plug it back in, and download his power at 12 cents per kWh. This will be possible with smart meters.

Not only is the smart meter the talk of the utility world; so is the smart grid. As government programs help redesign and reconstruct the national grid, many conservation advancements will be integrated into the system. Experts all acknowledge that the present grid system is antiquated. As it is rebuilt, new energy transmission ideas will be incorporated with the effect of conserving energy. This will not be cheap, and without a doubt ratepayers will see the cost reflected in their utility bills. Added expense is always painful, but on the plus side, these innovations will educate us and make us more conscious than ever of the need for conservation.

Smart meters will save each user money and energy, but what about the utility of wind turbines? A major criticism of wind turbines is that they cannot be controlled and do not necessarily generate electricity when the grid manager needs it. There may be no wind at 4:00 P.M. on a hot summer afternoon when the grid manager is desperate for power. Thus wind turbines cannot be counted on to meet peak demand. Utility companies spend billions of dollars to meet peak demand by having plants in reserve. These plants, mainly natural gas these days, stand idly by waiting to meet demand when the wind turbines do not respond. This uncertainty is frustrating to a utility company.

What can smart meters do about this dilemma? Utility companies expect that the smart meters will rearrange people's electricity use, and those high-priced load peaks will become hills. In other words, if we can level electrical use over a twenty-four-hour period, utility companies can eliminate costly, heat-producing, polluting plants as well as the energy lost in ramping up and down their auxiliary plants. Thus for the wind energy industry, smart meters will be advantageous. A constant complaint about wind installations is that the wind blows at night, and utility companies do not want night power. They consider it next to worthless because demand is low. However, if customer nighttime demand increases, that wind power will become much more valuable to both wind farms owners and the utility company.[23]

Subsidize conservation. Back in 1975 three young Environmental Defense Fund staffers, working in a vacated Berkeley fraternity house, came up with a novel idea. The Pacific Gas and Electric Company was gearing up to build new generating plants, using coal, oil, and nuclear fuel. The company projected steady growth in consumption, and the plants had to be built to meet demand. But the EDF staffers had a different solution. Rather than continue to meet projected demand, PG&E should work to *decrease* consumption for its product. Increased demand would be met by conservation measures, not increased capacity. To the utility company's planners and lawyers such an idea was certainly counterintuitive, but when the proposed $5 billion Allen-Warner Valley power plant was abandoned by PG&E and Southern California Edison, it was a victory for the soft energy path. PG&E would foster conservation through home energy checks, encouragement of solar water heaters and passive solar architecture, and such items as efficient lightbulbs. Who would pay for this? Again the EDF staffers had a solution. PG&E would recoup its conservation costs by adding them as a cost of doing business when applying to the California Public Utility Commission. Furthermore, the utility company could make a profit through conservation. Zach Willey, one of the EDF staffers, stated that a utility company should be able to make money by *not* producing electricity. Willey argued: "If PG&E makes the investment, they should be able to put it into the rate base and earn a profit on it, just like any other investment they make."[24]

This new environmental approach meant that California built no huge power plants in the 1980s, instead relying on renewable energy and proactive conservation. Unfortunately, the concept that a utility company can make money by not producing electricity has had only limited success. When a utility company allows you to buy blocks of wind power, that is added cost for the consumer. Utility companies need to foster conservation through home insulation programs, free low-energy lightbulbs, installation of smart meters, and all sorts of energy-saving devices. They should be in the business of conservation of electricity as well as production. In Jackson Hole the utility company provided customers with a dozen free low-energy lightbulbs and with water-saver shower heads. Presumably the utility's conservation costs have been added to the rate structure. Such conservation programs should be encouraged across the country.

Individual Subsidies of Green Power

Until recently most individual consumers were powerless with regard to the fuel that produced their home electricity. Those who paid attention were not necessarily pleased that coal was probably the culprit. Now there are options. At least 750 utilities across the country offer the consumer the option of paying a premium for green power. More than six hundred thousand households have signed up for the programs, believing that their premium costs are lessening global warming and reducing greenhouse gases.

How is it working? No doubt most of the utilities are funneling the money to the intended purpose. Unfortunately, as Stephen Smith of the Southern Alliance for Clean Energy notes, some large utilities are "preying on people's good will." They are spending more than half the money on advertising. Duke Energy's "GoGreen" program saw less than 18 percent of the premiums go to producing green power and 48 percent to marketing the program. The Florida Power and Light Company record is not much better. Its "Sunshine Energy" program had over 38,000 customers paying a premium of $9.75 each month. Less than half of the money made it to the growth of renewable energy. In a regulatory hearing in July 2008, Florida Public Service Commissioner Nathan Skop killed the program, stating that it was "all about PR and of little substance.[25]

Hence, as with choosing stand-alone turbines, if we want to contribute to a better world by buying blocks of green energy, the watchwords are: let the buyer beware. Some utilities, energy companies, and individuals are primed to take advantage of our guilt and naivete about energy. Most conservation programs may be legitimate, but due diligence is necessary.

Even if individual efforts work, the federal government must step up to the plate. It has already done so with regard to subsidies for energy production. It must also initiate and/or continue conservation programs through subsidies, public programs, and above all tax credits. Many of us will do the right thing without compensation, but incentives help a lot too. Poorer Americans simply cannot afford the costs involved with energy efficiency, while some people who are able to afford it will not act without a significant economic carrot.

Progressing into the second decade of the new century, we now recognize the energy problem. We cannot shake off awareness that the world has serious energy and pollution issues, and we Americans are at the center of these. As the comic-strip character Pogo memorably declared: "We have met the enemy and he is us." This chapter has presented an admittedly cursory view of the dilemma and described some of the ways to confront it. No energy source alone will solve the problem, but wind energy now takes its place as a contributor.

CHAPTER 10

Living with the Wind and the Turbines

The wind goeth towards the south, and turneth about unto the
north, it whirleth about continually, and the wind returneth again
according to his circuits.
> —Ecclesiastes 1:6

One thing about this country I don't like—and only one. The wind.
> —Owen Wister, 1885

The buffalo were a gift. The wind is a gift.
> —Rosebud Sioux, 2008

Wind now has a reputation as a useful natural resource, but it was not
always seen that way. The central figure in Dorothy Scarborough's 1925
novel *The Wind* is Letty, a refined southern woman who follows her new
husband to Sweetwater, Texas, in the 1880s. The book focuses on the
winter of 1886–87, when blizzards, drought, and the ever-present wind
changed the economy of the cattle-raising business forever. Letty did not
do well in that environment. Toward the end of the book the wind becomes a living entity, driving the vulnerable Letty crazy to the point that
she commits murder and then flees across the prairie "like a leaf blown in

a gale, borne along in the force of the wind that was at last to have its way with her."[1] She would not come back alive.

For years the town of Sweetwater suppressed the novel, and there is a legend of at least one book-burning in the town square. Now Sweetwater is the center of a great wind energy development. The largest wind farm in the nation—Horse Hollow—is located just to the south of town. Now Dorothy Scarborough's novel is regularly enacted as a play. A wind turbine might have made a difference for Letty. With electricity she would have had a potent ally to combat dust, dirt, drought, heat and cold; perhaps she could have coped with the grating environment.

During the Depression of the 1930s the wind was again the culprit in human disaster. It was viewed as the malicious force that created the Dust Bowl as it sifted the topsoil from the plowed plains, swirled it skyward, and carried the soil across the continent in a great dust cloud, to deposit in where it was not needed. Again, the wind caused mental anguish. A Northern Plains wife recalled that "my husband stood [the wind] for two months, watching our farm blow away in the winds, listening to that awful whistling, and then one day he hauled open the door and fought his way out onto the porch, yelling and screaming so hard it broke my heart."[2] Like Letty, he disappeared into the gale and was never seen again.

Such tragic stories exist for every era and in every part of the world; there are more than two hundred *named* winds worldwide.[3] Most come with plenty of legends, some of which touch on the wind as a catalyst for broken romance, mental illness, and suicide. Occasionally, people can literally be swept away. Marq de Villiers tells that when he was a child in South Africa his life was almost cut short. "In the grip of the gale, the child skidded across the grass until he landed with a crack against the metal railings that were all that prevented him from being hurled into the ocean. . . . It was there that my mother came, and fetched me away, and tried to still my terror with her beating heart."[4] The effect of the wind on most of us is rather less dramatic, more in the range of destroying um-brellas and coating clean cars with dust. As a fly fisherman I have cursed the wind when it conspired to prevent a graceful, accurate cast. When asked about Wyoming, the novelist Owen Wister said: "One thing about this country I don't like—and only one. The wind."[5]

Wister was speaking of an annoyance, but the wind can be a killer. One need not be reminded of a tornado's destructiveness. On mountain lakes, such as Yellowstone Lake, the wind can create waves of fearful proportions in a matter of minutes. In winter, and even at times in summer, a cold wind can sweep in, causing frostbite or worse. Temperature is important, but every outdoor person pays just as much attention to the wind chill forecast. In the American West wind is an ever-present force.

What can be said in praise of the wind? Probably most important, it brings change. If the sky is gray, the wind will turn it sunny again. It transports moisture, so absolutely essential for life. When the weather is unbearably hot, sooner or later the wind will bring cooler air. We generally appreciate a breeze as gentle and benevolent, making outdoor tasks easier by offering a natural cooling system and creating the dulcet sounds of nature we enjoy. If you are in mosquito country a good wind will blow the pests away. If you are a sailor, the wind is your partner. Your skill in using that partner yields a satisfying experience. And there is nothing worse than being becalmed. If you are in arid country and the rains fail, the wind will power a windmill, providing moisture from the earth rather than the sky. In short, we are ambivalent toward wind.

The Intimacy of Windmills

In time almost every American who travels will encounter wind turbines, and because they are so obtrusive, so large and imposing, the traveler will have an opinion. In contrast, the water-pumping windmills still scattered all over the West are intimate and familiar. My brother and I created a boyhood game from the windmills as my father drove our car from San Francisco to Modesto and then on to the Sierra Nevada. My brother would count windmills on one side of the highway, while I would count on the other. I do not remember who won those games, but I do recall that we would each tally more than one hundred. Plenty of them still stand, especially in the Altamont Pass area, but they are dwarfed by the commercial turbines.

We loved those windmills. Although we could not have expressed it at the time, they did represent pleasing additions to the landscape. They seemed to belong there, not overwhelming the rolling green hills. They represented modest intermediate technology at its finest. They have become icons of a bucolic, rural life, which even then a couple of urban kids admired.[6] Young, innocent thoughts aside, these small, understandable machines occupied space that was psychologically and physically apart from the city. Geographer Yi-Fu Tuan believes that the ideal landscape is between the city and wilderness, a place historian Leo Marx calls the "middle landscape." In our agrarian myth, this is the fertile, productive, yet natural place between the city and raw wilderness. It represents "man poised between the polarities of city and wilderness."[7] The traditional windmill harmonizes with this ideal space.

Wind Turbines Are Out There

It would be refreshing to believe that the modern turbines of today could fit into this space, simply a replacement and refinement on the old water-pumpers. In a way they are, for both employ the wind to spin blades, which in turn accomplish work. But there the similarities end. The difference is less a matter of appearance than of scale. The three-bladed Danish wind-mills are not ugly in themselves, but observers are often appalled by their size, with towers reaching well above 250 feet and blades close to 400 feet. The truths that they can be noisy, that they exist in clusters, and that they *move* all diminish their natural setting, leaving the perplexed observer to contemplate military-industrial metaphors—if not an invading army, at least an ominous industrial intrusion.

Wind developers at Altamont Pass hoped the green belt advocates trying to halt the ever-growing suburbs could be persuaded that the thousands of wind turbines actually protected the land from suburban development and that the turbines had a very small footprint. It did not work. Open space advocates bemoaned the arrival of the turbines and saw them as a utilitarian, semi-urban industrial landscape, no better than suburban sprawl.[8]

As with our ambivalence toward the wind itself, there is no universal attitude toward wind generators. The work of two academics directly addresses public opinion. Robert Thayer and his team from the University of California, Davis, identified the NIMBY response if a large wind turbine was within four miles of the viewer. More than four miles would be acceptable. Maarten Wolsink, a researcher in the Netherlands, found that the NIMBY phenomenon came primarily from *seeing* the turbines on the landscape. Wolsink believes people may object to noise, shadow flicker, and bird death, but visual intrusion remains the root cause of opposition.[9] Thayer and Wolsink both provide excellent data on individuals' responses to wind turbines.

Opinions range from love to hate, with many variations in between, and scientists have difficulty dealing with such subjectivity. When the National Research Council's Committee on Environmental Impacts of Wind Energy Projects recently published their findings, they shied away from the most important subject. In the chapter examining the impacts of wind development on human beings they decided to pass over "all possible human impacts from wind energy projects." They did not include such aspects as "significant social impacts on community cohesion, sometimes exacerbated by differences in community make-up," meaning differences in values, wealth, and age. They also would not address "psychological impacts—positive as well as negative—that can arise in confronting a controversial project."[10] Clearly this was too subjective a topic for the scientists to address and reach any sort of consensus; the subject does not lend itself to meaningful statistics.

Yet they did address one key impact on human beings: the fact that those individuals and families who suffer negative visual or noise effects from the turbines live too close to them. This is not the fault of the homeowner, for in most cases the home was erected before the wind turbines arrived. Usually it is attributable to local government regulations, which often allow setbacks of only 1,000 feet. Significantly, in their study the National Resource Council's wind committee observed that "the most significant impacts are likely to occur within 3 miles of the project, with impacts possible from sensitive viewing areas up to 8 miles from the projects."[11] One might expect that this would preclude setbacks of less than at least a mile. But the industry prefers setbacks measured in feet rather than miles.

When wind developers win, as they often do, controversy ensues. All over the country local hearings have turned into shouting matches. People protest that their lives have been ruined by the huge machines raking energy from the sky. Tony Moyer and his wife are typical of thousands. Wanting to live and raise their children in the country, they bought a 35-acre farm within the town limits of Empire, Wisconsin. Their new neighbors are the forty-four turbines of the Cedar Ridge Wind Farm. Three of the turbines are located within a quarter mile (1,320 feet) of their home and twelve are within view. Although the wind developer made promises, the noise and vibrations are more than disturbing. The Moyers cannot open their windows at night, for outside, says Mrs. Moyer, "it truly sounds like I am at O'Hare airport." Even the interior of their home is seriously impacted. She says she has not had a decent night's rest since the wind turbines were turned on: "Each night I need to take a pill to sleep. Sometimes that doesn't even help."

Her children also have trouble sleeping. As if that were not enough, late in the afternoon, around 6 P.M., "our home has been turned into a disco with the shadow from the blades." She vows that when possible she and the family will move "to somewhere peaceful."[12]

Is Mrs. Moyer exaggerating? Does she have some alternative motive? Does she wish to lower her property tax or perhaps sue the wind energy company? Is she a crabby person who thrives on complaining? These are possibilities, but I believe she and her family have genuinely suffered because of poor decisions by the township government and inaccurate information provided by the Cedar Ridge Wind Farm developer. The siting of wind turbines within a quarter mile of homes is simply too close. Often the audio testing by the developer is done during the day, when atmospheric conditions are such that quiet prevails. I have stood next to commercial turbines during the middle of the day in Oklahoma and wondered what the problem is. The problem is that conditions change at night, with wind direction and weather factors. There is no ambient noise to cover turbine hum, and atmospheric conditions often magnify the moving sound.

Wind energy developers religiously test the average wind speed of a potential wind development for at least a full year. When it comes to whether they can make money, they collect plenty of data. The industry should give the same attention to noise, gathering information on anticipated noise and vibration intrusions. Some may claim this cannot be done without having the turbines in place, but surely accurate predictions can be made. Also, comparative work can be performed with other wind farms with similar terrain, prevailing winds, and atmospheric conditions. The reality is that once the turbines are in place, little can be done.

Wind Turbine Syndrome

One thing we know about noise is that it affects different people in different ways. I like music or the television sound to be higher than my wife wants it. Part of our difference is that her hearing is more acute than mine, and part is subjective preference. The perception of noise depends in part on the individual. Some people protest at the persistent whoosh of a turbine, while others may find it soothing. I believe the latter are far fewer than the former. I have read so many sad, angry, or passionate letters about wind farm noise, mostly occurring at night, that I am convinced Mrs. Moyer's complaints have validity and mirror widespread troubles.

Another person who takes this point of view is Nina Pierpont. She is a medical doctor, although not an audio specialist. She practices in Franklin County, which she describes as "the poorest in New York State." When the wind developers came to her county and others in upstate New York, she took an interest, particularly in human health effects. After the turbines were up she found repeated health complaints from people living in close proximity to the wind installations. Eventually she identified the symptoms and gave them a name:

> Wind Turbine Syndrome is the clinical name I have given to the constellation of symptoms experienced by many (though not all) people who find themselves living near industrial wind turbines: sleep problems (insomnia), headaches, dizziness, unsteadiness, nausea, exhaustion, anxiety, anger, irritability, depression, memory loss, eye problems, problems with concentration and learning, tinnitus (ringing in the ears). As industrial windplants proliferate close to people's homes and anywhere else people regularly congregate (schools, nursing homes, places of business, etc.), Wind Turbine Syndrome likely will become an industrial plague.[14]

What are we to make of this? Wind energy salespeople slander the doctor's work, claiming she has no evidence and probably believes wind turbines caused mad cow disease. Such dismissals only hurt the industry. Pierpont has written a book on the subject, and the work has had a number of peer reviews from English, Canadian, and American physicians.[15] People have written as advocates of her theory. Rather typical is a letter from a high school teacher of twenty-seven years, who teaches science classes. She has lived on a fifteen-acre farm for nineteen years. In November 2007 turbines began spinning on adjacent property. Immediately she had trouble sleeping and soon experienced a humming in her head just behind the ears. She had trouble remembering the names of her students and took much longer to prepare her lesson plans. She became emotional and cried a great deal. She could not sleep. To the wind energy company's credit, it has tried to resolve her problems, providing a place in town away from the turbines. But the hum and the ringing in her ears have not gone away.

Perhaps our high school teacher is the exception rather than the rule. Others live close to wind turbines without such negative effects. Yet the evidence is mounting. In March 2009, Dr. Michael A. Nissenbaum, a radiologist at the Northern Maine Medical Center, decided to interview fifteen people who lived within 3,500 feet of the wind installation at Mars Hill, Maine. His findings, presented to the Maine Medical Association, were alarming. All fifteen of those interviewed reported that their quality of life had been negatively affected by the turbines. As if to substantiate Pierpont's theories, "residents all expressed new or increased feelings of stress, anger, irritability, depression, anxiety, and hopelessness." These feelings, noted Nissenbaum, were so strong and so high "that despite the small number, and the lack of controls and tests of statistical significance, they jump out at physicians as obviously being significant."[16]

Pierpont's work and Nissenbaum's interviews do not represent hard and fast evidence, but they certainly give pause. There is also other anecdotal evidence that these turbines affect individuals in sometimes violent ways. In Eastland County, Texas, sixty-seven-year-old Ray Johnson ended up behind bars when he was caught exercising his habit of shooting his rifle at the Silver Star I Wind Complex, run by BP. He was also charged with assaulting an employee. His criminal mischief charge came from emptying his rifle at Turbine No. 6, doing at least $20,000 worth of damage. Is Johnson mentally disturbed, or is this a case of wind turbine rage?

In England, Arnold Wilkins, a neuropsychology professor at the University of Essex, believes that the flicker of turbine blades can cause epileptic seizures. He calls for guidelines. Perhaps more seriously, inmates at England's Whitemoor Prison have suffered psychological disturbances as a result of an adjacent wind turbine. The noise may have contributed to five suicides within the year. Alan Devlin, an employee, claims that "the risk is unacceptable for a captive audience, some with psychological problems, unable to leave the area."[17]

Pierpont recommends that the giant commercial turbines be installed no closer than 2 kilometers from an existing home. Other researchers and residents who live close to the turbines agree. We have no national or state standards on setbacks, but it is time for discussion of these. When I first started studying the NIMBY response to turbines I was convinced that viewshed issues were at the heart of people's response. Now I realize that the noise effects are more significant, particularly because residents do not anticipate such strong reactions until the turbines are up and running—by which time, of course, it is almost impossible to perform meaningful mitigation.

The Native American View

When we examine mental and physical health issues associated with wind turbines, we need to put them in perspective. They are serious and widespread but not nearly so grave as those associated with other forms of energy production. American Indian tribes understand the downside of energy production. They have been intimately involved with the mining of coal and uranium. Although the tribes own 2 percent of the nation's land base, they control 30 percent of the coal reserves, 37 percent of the uranium, and 10 percent of onshore natural gas deposits.[18] They also have a rich untapped wind resource. The coal and uranium reserves have been a mixed blessing. While bringing income and jobs to the reservations, they have also contributed to devastating health problems.

The story of Navajo uranium miners is well known. In the 1950s and 1960s they "worked in dusty mine shafts, eating their lunch there, drinking water from sources inside the mine, and returning home to their families wearing dust-covered radioactive clothing." They were given no clue that the mine would be hazardous to their health. Their children regularly played in the uranium tailings. In the case of the 150 miners at the Shiprock mine, by 1975, 18 Navajo miners had died of cancer, by 1980 an additional 20 were dead of the same cause, and another 95 had contracted serious respiratory ailments and cancers: that comes to 133 out of 150.[19]

Such health statistics are astounding, and figures are not much better in the coal fields. When it came to mining and energy, as historian Colleen O'Neill has expressed, the Navajo workers had to weigh "jobs and royalties against their own health."[20] Of course the Navajo people do not have a monopoly on health issues related to energy production. Throughout the West, peoples living on reservations face difficult choices that are beyond the American mainstream. Often they must endanger their lives to make a living.

Wind Energy: Solving a Dilemma

On the Navajo Reservation interest in wind energy has come about as an alternative to more of the same traditional energy production. The proposed Desert Rock coal-fired power plant excited plenty of opposition because of air quality and global warming issues but also because the Navajo people are tired of having their lands ravaged and their health devastated. As Indian historian Don Fixico states, "Tribal love for the Earth is strong and it is the fiber of the souls of the native peoples."[21] At the time of writing the fight has gone on for over five years, but the company sponsors in Houston and New York are losing heart, especially with the difficulty of raising some $4 billion dollars in a difficult economy.

Some younger Navajo leaders, in tune with environmental issues as well as Navajo spiritual beliefs, saw that if they were to defeat Desert Rock, they must present an alternative energy plan. To do so they spearheaded the report "Energy and Economic Alternatives to Desert Rock Energy Project." The report found that an equivalent amount of energy as was proposed at Desert Rock could be produced through a combination of wind and solar power and natural gas.[22]

The tide quickly turned. Wind energy offered a practical solution much more in keeping with the cultural and spiritual beliefs of the Navajos. A little over two months after distribution of the report, Joseph Kennedy II's small plane touched down at Window Rock. On March 27, 2008, the Diné Power Authority and Joseph Kennedy II, the chairman and president of Boston-based Citizens Energy Corporation, signed a joint venture agreement to build and operate a 500 MW wind farm. Kennedy announced that "this will be an exciting project that can also put very significant amounts of money back into this reservation. I'm very hopeful," he continued, "that we can create a sort of all-winners opportunity."[23] In essence, both groups agree on the worthiness of the project, with the Navajos providing the site and Citizens Energy providing the capital. The Navajo Nation expects to gain $60 to $100 million over the life of the project, the health of the people should improve, jobs will be created, and the air will be pure.[24] If the project proves successful, others could follow. With its millions of acres of open land, and with a potential of 15,000 MW of wind power, the possibilities are endless.

On the Northern Plains

Farther north in the "wind corridor" of the United States, there is ample evidence of wind energy potential on Indian reservations. When Anglo Americans invaded and conquered the Great Plains in the nineteenth century the government placed the surviving Native Americans on reservations. Government officials hoped they would take up farming, the children would receive an Anglo-oriented education, and the tribal groups would be assimilated. The policy failed. The Indians did not blend into the larger society. Although much of their surface land is marginal for farming, beneath the soil lay valuable resources. Above the land flows the ever-present wind, so long unappreciated.

Native Americans are now poised to participate in the green economy of the future. Four Indian organizations—Honor the Earth, the Intertribal

Council on Utility Policy, Indigenous Environmental Network, and International Indian Treaty Council—lead the effort on the Northern Plains. Members believe that renewable energy opportunities will allow the tribes to break the cycle of choosing between economic development and preservation of people, culture, and land. In other words, wind turbines will break the toxic legacy left by fossil fuels and uranium development and stave off the threat of nuclear plants and storage facilities. With wind energy there will be no false choice between polluting energy development and dire poverty.[25] Quite obviously, the Indian organizations will choose wind energy, knowing that it is the best way to provide jobs and money, while preserving their culture and attachment to the land they treasure.

The hope is that the nine tribes of the Northern Plains can develop more than 200 gigawatts of potential wind power. The energy will power the reservation and will also be sold and exported to populated areas in the United States.[26] New transmission lines will be needed, as will private or government funding for turbine construction. The tribes would like Congress to review the subsidies presently paid to the coal, gas, oil, and nuclear industries and have these billions redirected to sustainable energy development. Such a massive infusion of cash to renewables is not likely, but with an administration sympathetic to both Indian interests and alternative energy, anything may be possible.

At the moment most of the plans for wind energy are still on the ground. However, Robert Gough of the Council on Utility Policy has been able to erect a 750 kW turbine on the Rosebud Reservation. The Department of Energy offered the tribe a 50 percent match for the project. Finding the other funding was not easy. Finally, the Department of Agriculture's Rural Utilities Service put up a loan for the remaining 50 percent. Thus, 100 percent of the project funding was underwritten by the federal government. With this successful pilot project, Gough expects to expand to 80 MW (10 MW clusters on eight reservations). Clearly funding is a great hurdle, but federal loan guarantees could get the planned projects into the air. At the moment reservation leaders and wind company executives are often at odds. As one Rosebud Sioux tribal council member said of a company representative: "He questions our mentality. I question his." Development will come, but Indian leaders are tired of disadvantageous contracts. Much mistrust remains, even though wind energy on reservations is certainly a win-win situation. As one Rosebud Sioux put it: "The buffalo were a gift. The wind is a gift."[27]

Some Indian tribes live on reservations that have plenty of wind. They need only the capital to realize its potential. Here Oglala Sioux workers raise a refurbished turbine to power a tribal radio station on the Pine Ridge Reservation in South Dakota. (Photo by Bob Gough, NREL.)

Tribal leaders are particularly concerned with potential impacts on wildlife (avian mortality) and the landscape. On reservations where even a two-story house is a rarity, a 400-foot turbine is jarring. The 750 kW turbine at Rosebud is the tallest structure in South Dakota, save a few communication towers.[28] Multiplying the one 750 kW turbine by one hundred larger ones is a serious matter. There may be a strong NIMBY response to large-scale commercial wind parks on Indian land, but the health and economic advantages will make development attractive.

Education Is the Key

Back in the 1980s the appearance of a wind farm was bound to elicit a response of "What is that?" Now most Americans recognize such installations, although many still do not know what they do. If the public is to embrace this new energy world, people need a basic knowledge of electricity production, the grid system, and ways of conservation. School age children are one target audience: they are young and open to change. Recognizing this fact the National Renewable Energy Laboratory established a program called "Wind for Schools."[29]

The program identified the key states as Idaho, Kansas, Colorado, Montana, Nebraska, and South Dakota: states with significant but largely untapped wind resources. It calls for installation of small wind turbines at elementary, middle, and high schools. State universities provide a co-ordinator, and engineering students learn skills by installing the turbines. Such major wind developers as Babcock and Brown, Iberdrola Renewables, Horizon Wind Energy, and Tradewind Energy have all supported the program through encouragement and monetary donations. Small turbine manufacturers, such as Southwest Windpower, provide turbines. Obviously the objective of the program is education and enthusiasm, not a large revenue stream. Students learn about the science and technology behind wind energy by monitoring the turbine's output in real time on their computer screens.

Program coordinators report a certain fervor. Ruth Douglas Miller, the engineering professor who leads the effort at Kansas State University, reported that at one elementary school "the students were just ecstatic to have [the turbine] going. There was really a lot of excitement."[30] Here was a whirligig that was more than a mere toy. For full acceptance of wind turbines, both the government and trade organizations such as the American Wind Energy Association must sponsor such educational forums and

Faculty and students break ground for a wind turbine at Sanborn Central School in Forestburg, South Dakota. This is a Department of Energy–funded project. (NREL photo.)

demonstrations. With a minimum of educational effort the wind can be transformed in a child's mind from a negative to a positive force. The children were happy to see the wind blow because it turned their windmill, creating electricity that they use and appreciate. This young experience will stay with them for life, and it is doubtful that they will curse the wind or see it as anything but an affirmative force.

Any study of individual and societal perceptions of wind energy must conclude that human beings find the new industrial turbines a jarring makeover of the landscape. Sometimes we welcome change, but when it is forced upon us, it can cause unwanted consequences. In this chapter we have seen some of the less desirable consequences of the remarkable growth of wind power. No matter how desirable wind is as a clean source of electricity, sometimes it brings a bitter harvest, fraught with human disorders reaching the point of drug dependency. No one wants that. There are times when the wind energy industry must show flexibility, fairness, and compassion.

Conclusion

Over the past thirty years there have been plenty of critics of wind energy. They proclaim that it does not work without significant subsidies, and even if it did work, the sacrifice of scenic land or seashore is simply not worth the meager amount of electricity produced. We now know that the critics are wrong. Government subsidies are modest, indeed, compared with those for fossil fuels and nuclear power. Above all, wind energy transforms nature's energy to electricity in a most gentle way: one in which the planet and its inhabitants are not endangered by heat, pollution, toxicity, or depletion of irreplaceable natural resources. That sentence says a lot. Throughout our utility history such environmental benefits were not figured into the cost of energy. Now environment must trump economics. Translated, this means we should either use less electricity or use more benignly produced electricity. Logically, the nation should embrace the former option. We must conserve, and enterprises and individuals refusing to do so should pay dearly for wasting a precious commodity. Unfortunately, many people consider it their right to consume as much as they like. The result is that national conservation efforts have not been successful, from Jimmy Carter's presidency to the present. Progress has been made in education, but not to the point where we collectively use less electricity per capita and shrink our national footprint upon the resources of this earth. We should, it would seem, continue to produce more electrical energy, and wind energy is the most benign, earth-loving way to achieve that objective.

The Long View on Reliability

One of the denunciations of wind energy is that it is unreliable. Certainly there is truth to such criticism in the short run. However, we need to think in terms of years, decades, and centuries to realize that wind energy is the most reliable of all means of power production. One hundred years from now, Americans will not be relying on fossil fuels for—at our present consumption rates—they will be gone, except perhaps for coal. Just what technology will be devised I cannot say, but I do know that the wind will be with us. As long as the sun shines, the wind will blow. That is reliability no scientist or engineer can dispute.

Wind is more reliable than hydro power, a significant renewable resource. Particularly in the American West the wind is ever present, but water is not. Climatologist Connie Woodhouse assures us that over hundreds of years drought "is a part of natural climate variability in this part of the world [the Southern Plains]."[1] Among the consequences of drought are that sand dunes emerge, crops shrivel, livestock die, and farm families pack their belongings to seek a livelihood elsewhere. The land can become uninhabitable for human beings. Drought can also mean that production of electricity through water power is significantly reduced: it becomes unreliable. Such a utility as Idaho Power can produce 50 to 60 percent of its power from dams, mainly on the Snake River. But a low snowpack year can reduce hydro power output to 20 to 30 percent. Moreover, it should be added, traditional nuclear, natural gas, and coal power plants require millions of gallons of water. Wind turbines are more flexible: they require no water.

Utility managers scoff at such attempts to prove the reliability of wind energy. They do not have the luxury of thinking and planning decades ahead. Their time frame is the next twenty-four hours. Can they rely on those turbines to produce peak load electricity for a hundred thousand homes on a 110-degree day? Realistically, they cannot. Sometimes mother nature does not cooperate. This is the Achilles' heel of wind energy, and the reason why it will always be a supplementary power source. However, traditional fossil fuel sources are not always dependable. In 2004 New England suffered a severe cold snap. The grid, called the Independent System Operator (ISO), was stressed, but ISO load manager Stephen Whitley was not worried since he had 10,000 megawatts of gas-fired power ready to fire up. But when he called on the natural gas plants to come on line or increase their output, he was shocked when plant operators said they were ready to go but could not purchase natural gas at any price. It was not available. Fortunately, the cold snap eased and Whitley averted the crisis.[2] Other cases come to mind when fossil fuels were unreliable or at least insufficient to meet a crisis, as in California.

Since we know wind energy is unreliable, is there any way we can manage the inherent uncertainties of wind energy? Fortunately, there is. Today a number of companies specialize in forecasting. Of course wind forecasting must be performed before any utility-size project is undertaken. However, of more interest is short-term wind forecasting once the wind farm is operational. Managers and grid managers need to know the amount of wind power that will be produced each hour in order to balance the expected load. This is the job of such companies as Seattle-based 3TIER. The company uses sophisticated statistical models to forecast wind speed in a specific wind regime. Among many reasons for such forecasting surely the most important is "to determine the amount of power to be produced from conventional power plants" so as to "balance the intermittent output of the renewable resources."[3] The idea, of course, is to maximize use of wind power and augment it with conventional sources, so as to produce the needed load and waste nothing. If the forecasters can perfect their weather and wind predictions, they will resolve one of the major drawbacks of wind energy.

An Exciting Field

For an environmental historian the study of wind energy is indeed exciting. Today throughout the United States we are building a new primary energy source based on natural processes. The wind has been employed since the dawn of humankind. Medieval people refined its use through the pre-industrial age, when power and work were the preserve of wind, water, sun, animals, and humans. Aside from hydro power, however, the inventors and scientists of industrial times had little appreciation for any of these power sources; fossil fuels were ascendant. Now the tide is turning and nonrenewable fuels are going down the path of the dinosaur. Even if such sources were abundant, we realize that their continued use may make the world uninhabitable. In the midst of the energy dilemma wind energy has emerged. Still small in its overall impact, it has the potential to provide a significant portion of our energy needs without harm to life. Like many others, I have been skeptical when political leaders or experts claim that our environmental problems can be solved through technology. But wind energy gives me hope.

That hope is doubled when we look at the economic front of manufacturing and jobs. Each day wind energy growth is creating what have been so-called green-collar jobs. Oil patch jobs in Wyoming, for instance, are disappearing as production declines. Natural gas wells are picking up the slack, but they will not produce forever. In the meantime, companies are developing a resource that will be on earth in abundance for as long as we wish to harvest it. Out in Fort Stockton, West Texas, Doug May proclaims the benefits of nearby wind farms in terms of jobs. Although there is a difference between an oil pump jack and a wind turbine, for a skilled mechanic much is interchangeable. Again, the oil and gas will run out; the wind will not. May understands that wind turbines guarantee the continued existence, and even prosperity, of his town.[4] The prominent American West historian Elliott West expresses it well: "How lovely it would be if the Southern Plains should regain some economic footing by capturing one more abundant form of energy, converting it as so often before into usable power, and exporting it to those in need, in this case the power-starved residents of the Pacific Coast who look down from airliners into the apparent emptiness of West Texas and who drive across it as fast as possible, goggling at the country's sparseness and complaining about being buffeted by the distilled essence of this country's long story of restless movement—the wind."[5]

Because of automation and technology, we cannot count on the wind farms to create large numbers of new jobs. After construction, wind farms will employ only about one to two people per 20 MW. A century ago railroad yards, shipping ports, telephone switching centers, and even farms were swarming with workers; today they are eerily empty.[6] Wind installations are no different. However, wind energy is also creating high-paying manufacturing jobs. In 2004 only 30 percent of wind energy components were made in the United States. In early 2008 that figure had increased to 50 percent. New factories are being built. The prominent Danish company Vestas is building a $245 million facility in Colorado to produce nine hundred towers per year. It will employ about four hundred people. This plant is in addition to a blade factory already in production in Windsor, Colorado. Siemans Energy will enlarge its blade factory in Fort Madison, Iowa, adding two hundred new workers.[7] This sort of growth is evident throughout the nation. Green energy is producing good manufacturing jobs.

Butte, Montana, a town that has known hard times ever since the Anaconda copper mines closed, will have a blade factory run by Fuhrlander Company, a growing wind turbine manufacturer.[8] Every wind energy company, particularly General Electric, has orders for more turbines than it can handle. A significant backlog is good news for a suffering economy.

Wind Energy Growth

We have every reason to assume the boom will continue as long as federal and state governments support renewable energy. Rosy predictions are always risky, but when former U.S. Secretary of Energy Samuel Bodman addressed the Washington International Renewable Energy Conference on March 4, 2008, he expressed the vision that wind should provide 20 percent or more of the nation's generation capacity. More recently, Secretary of the Interior Ken Salazar easily outdistanced Bodman when he announced at a public hearing in Atlantic City, New Jersey, that offshore East Coast wind installations could produce 1,000,000 MW of electricity, making it possible to replace 3,000 coal plants.[9] This would require 500,000 wind turbines of 2MW capacity. Even the wildest advocate of wind energy would admit that Salazar is off the mark.

Such wild predictions by a prominent government official are unwarranted, especially in an industry that has long suffered from hyperbole. Yet the industry and renewable energy are on a tremendous growth curve. In 2007 the United States installed 5,240 MW of new wind power, the fastest growing wind power capacity in the world. The pace has continued. According to *Wind Today*, between July and September 2008 workers completed twenty-four wind farms, representing 1,433 MW of capacity. New construction as of September 2008 involved eighty-six projects with a combined megawatt capacity of 8,663.[10] In California Governor Arnold Schwarzenegger issued an executive order that by 2016 the state must generate 33 percent of its electrical production from renewable resources. This is a remarkable challenge and a staggering amount of electricity. Probably at least 75 percent of the 33 percent will come from wind turbines. Of course the governor's executive order represents a goal, not a mandate, yet the state utilities commission must take action with this goal in mind. Although not as daring as California, states throughout the country have embraced renewable energy, and clearly a new awareness has gripped the American people.

Once More a World Leader

That awareness has translated into growth that has catapulted the United States back into the world leadership. In the 1990s Germany surpassed us in wind energy capacity through a number of creative stimuli, but the remarkable growth spurt of the last few years has brought the United States into the number one position with 25,000 MW of capacity.[11] If the United States keeps on course with all the wind installation plans, the gap will increase in 2010 and beyond. Wind is the power source of choice these days and represents about 30 percent of all installed electrical additions in the nation.[12]

The interest in wind energy can be measured in more than turbine capacity. The 2009 American Wind Energy Annual Conference in Chicago attracted about fifteen thousand attendees as well as twelve hundred exhibitors. The conference was 65 percent bigger that the 2008 conference in Houston, and the Houston meeting was 85 percent larger than the 2007 conference.[13] All this is a far cry from the dedicated small group that gathered in San Francisco in 1987.

Where to Put the Turbines

If the growth in the industry continues, a major question will be: where will we put all these turbines? First, of course, we must have the wind resource, which probably eliminates about a third of the nation. Second, we run into conflicting land use. Not everyone wants wind turbines, even if they bring money and help the environment; perhaps another one-third of the nation opposes having hundreds of turbines. Some years ago a brilliant landscape planner named Ian McHarg wrote *Design with Nature*. What he did was use colored overlays for the region in question to signify historic values, recreational values, residential values, water values, forest values, and scenic values.[14] If we were to test this kind of overlay system for the nation, I believe it would reveal where the large installations of turbines should be.

In view of what we already know, we may not need McHarg. The Great Plains of the United States constitute a vast region of low population that should one day bristle with huge wind turbines. It has the wind resource and is desperately in need of an economic shot in the arm. Most people associated with wind energy have identified the region's promise. As we have seen, one of the most prominent is T. Boone Pickens, who owns a 60,000-

acre ranch near Pampa in the Texas panhandle, and has announced a massive new wind energy power plant. For starters he has ordered 667 General Electric 1.5 MW turbines, which he will erect on or near his ranch. By 2014 Pickens would like to install another 2,000 turbines, giving his wind power station a capacity of 4,000 MW, at a cost of between $10 and $12 billion. The economic downturn notwithstanding, we might expect government subsidies and loans to get the Pickens plan under way.

Perhaps Pickens's biggest problem will be getting his electricity to the densely populated Dallas–Fort Worth region: the panhandle region is outside the Texas grid. If he cannot work something out with the Public Utility Commission, he plans to build a private transmission line.[15] This would be complicated since a private transmission line is unprecedented, but given the Pickens can-do spirit it may well happen.

Is the Pickens vision of about wind energy facilities stretching from northern Texas to the Canadian border fanciful? County after county has experienced out-migration. As mentioned earlier, sociologists Frank and Deborah Popper have recommended a buffalo commons for this vast plains region, consigning it to wildlife where people have failed to prosper. Now wind energy may represent a new dawn for struggling families and desperate towns. Huge wind farms might even attract tourists rather than repel them; visitors to Texas and Oklahoma can already pick up a Wind Power Trail brochure directing them to wind power sites, as part of changing the image of wind turbines.[16]

But we should not count on wealth or tourism to transform the country's use of renewable energy. It is too huge a project for private enterprise to fulfill. The job must have the total support of Washington politicians, who must rise above local interests and earmarks and morally internalize the green revolution as their first priority. Policy makers as well as citizens should embrace the ideas presented by Al Gore on July 17, 2008, for 100 percent clean electricity within ten years.[17] This is not a straightforward proposition. Most of the public and many politicians continue to embrace the status quo, and although they understand the dangers in over-reliance on carbon-based fuel, they insist on myopic baby steps, refusing to expand their vision past a few years. Furthermore, they need to support constructing many new transmission lines if we are to move the electricity from the wind turbines to the urban centers. In 2008 the nation elected leaders who embrace bold solutions, who are determined to break our reliance on combustion energy, and who are focused on future generations. Public policy will make or break the effort for a sustainable world; that is why Thomas Friedman warns us that "it is much more important to change your leaders than your lightbulbs."[18]

We have changed the leaders. Now we must support them.

Acceptance

Finally, a word regarding acceptance. Local opposition to wind turbines is a strong theme of this book. Sometimes, as in upstate New York, opposition has been based on alleged corruption by wind energy companies. Wind energy developers play on the economic weakness of poor communities and desperate families. Near the town of Burke, New York, the economic attractions of wind development have torn apart families to the point that sisters and brothers no longer talk to each other. The atmosphere has become so charged that in August 2008, New York attorney general Andrew Cuomo agreed to investigate what appears to be a culture of corruption and intimidation enhanced by wind developers.[19] Similar scenarios have bedeviled other regions.

The days of an oil patch mentality of greed and boom-bust cycles are about over. Most developers understand that it is in their best interest to operate openly and in good faith with the local community. More problematical is the question of landscape. Wind turbines placed in a pleasing agricultural, scenic, or historic landscape evoke anger and despair. At the heart of the issue is visual blight. Residents do not want to look at the turbines and are willing to fight wind development. Their wishes should be respected. Wind developers should take to heart geographer Martin Pasqualetti's advice: "If developers are to cultivate the promise of wind power, they should not intrude on favored (or even conspicuous) landscapes, regardless of the technical temptations these spots may offer."[20] The nation is large. Wind turbines do not have to go up where they are not wanted. We can expand the grid and put them where they are welcome.

There will always be opposition to change, but for some people the new turbines are evolving into icons of hope. Marketing author Seth Gordon believes icons tell "a story that validates our feelings and amplifies the way we look at the world."[21] The turbines certainly appropriate the landscape, but if we think of what they *do*, rather than how they *look*, we may come to think of the spinning blades as reassuring. Furthermore, they are not visually repulsive to everyone. Much of the Great Plains, the future growth

A spinning turbine can be a thing of beauty. Sculptor John Simms creates many revolving works. His *Dynamo in Bronze* does not generate electricity, but it could. (Courtesy of John Simms, Jackson, Wyoming.)

site, is unrelentingly level and without obvious visual appeal. Interstate 80 across eastern Wyoming, for instance, offers little scenic punctuation for four hundred miles. In his *Infrastructure: The Book of Everything for the Industrial Landscape*, Brian Hayes makes the case that juxtaposing industrial cylinders and spheres with natural land forms "adds visual interest to the composition."[22] Perhaps this applies to wind turbine rotors as well; we already see the old water-pumping windmills that way. He does not expect to win many converts to this point of view, but I would argue that the more than one hundred turbines along Interstate 80 between Laramie and Rawlins bring visual relief, and if they are not beautiful, they do evoke wonder and a certain curiosity. They are here, and they are multiplying, and we may even come to like them. Journalist Mary Jo Murphy calls on the turbines to "enjoy your moment . . . and blow a few kisses."[23] As we get serious about demanding clean energy, the turbines are an icon for those who contemplate our energy future.

A new administration came into office in January 2009 with change as its mandate. Certainly our future with regard to renewable energy is a high priority and one that will take money, time, and sacrifice. We are not accustomed to sacrifice. When the nation went to war a decade ago, our leaders asked us to do our part for the economy by going shopping. Thus far it is not much different for the environment. We are having a green revolution, but as environmental journalist Thomas Friedman states, the nation is not so much having a revolution as throwing a party: "We're actually having a green party. And it's a lot of fun. I get invited to all the parties. But in America, at least, it is mostly a costume party. It's all about *looking* green—and everyone's a winner. There are no losers."[24] Obviously, such a painless scenario will not accomplish the revolution. Friedman moves us ahead by time machine to 20 E.C.E. (Energy-Climate Era.) By this date Americans have accomplished the electrical revolution by an elaborate program he calls the smart grid. Wind energy will play a major part. Utilities will generate half their electricity from renewable sources. Southern California Edison, hypothesizes Friedman, will have "huge wind farms in Wyoming and Montana, and [will have] contracted with many smaller independents along the way. The Wyoming farm is so vast it is a tourist site, like the Hoover Dam, with turbines as far as the eye can see."[25] The picture he paints is a departure from our traditional understanding of the plains—until we view the way forward through the old pragmatic lens of harnessing regional potential, whereupon the scenario becomes futuristic but functional, selectively embracing a new kind of landscape to ensure a livable world.

Notes

Much of this story depended on a very recent addition to the researcher's store of information—the Internet. It has been extremely useful, and in spite of criticism regarding accuracy, it is the only source (aside from telephone calls and newspapers) for contemporary information. However, providing full citations for some Internet sources is difficult. If a note contains the full newspaper citation, it is usually because the printed copy was available. If the citation gives the date only, with—in some cases—no journalist cited or no page number, the Internet did not provide them.

More serious, perhaps, twenty or more citations are to the National Wind Watch site (www.wind-watch.org). This website describes itself as a "non-profit coalition for raising awareness of the shortcomings of industrial wind energy." It posts daily newspaper articles from all over the world. It was very useful, and it does not hide its opinions.

However, accessing an article is impossible. In October 2010, National Wind Watch posted the following note: "On August 9, 2010, National Wind Watch lost its main database of news stories, consisting of items back to August 1, 2008. These items are irretrievably gone from the site." Thus many of the sources referenced to this site are unavailable. Even those that are not within the time frame are difficult to find. I apologize. There are many advantages to our paperless world, but occasionally we may be too dependent on a somewhat untried technology.

Introduction

1. See Walter Prescott Webb, *The Great Frontier* (1951; reprint, Austin: University of Texas Press, 1975), 180–81.

2. *New York Times*, August 24, 2008, sec. 4.

3. See Wendy Williams and Robert Whitcomb, *Cape Wind: Money, Celebrity, Class, Politics, and the Battle for Our Energy Future on Nantucket Sound* (New York: Public Affairs, 2007). My guess is that by the time the present volume book is published, the Cape Wind project will be approved and under construction.

4. The classic examination is Leo Marx's *Machine in the Garden: Technology and the Pastoral in America* (New York: Oxford University Press, 1964); also see Andrew G. Kirk, *Counterculture Green: The Whole Earth Catalog and American Environmentalism* (Lawrence: University of Kansas Press, 2007), 8.

5. Watt is the measurement of electrical use at any moment. The measure, however, is in watt hours. If you use a 1,000-watt device for one hour, usage is a kilowatt hour (1 kWh). A megawatt equals one million (1,000,000) kilowatt hours. A gigawatt equals one billion (1,000,000,000) kilowatt hours. Thus 20 gigawatts is equivalent to 20,000 megawatts.

6. Reports in www.wind-watch.org/news/2009/01/10 and 2009/01/15.

7. Thomas Friedman, *Hot, Flat, and Crowded* (New York: Farrar, Straus and Giroux, 2008), 24.

Chapter 1. How Have We Used Wind and Electricity?

1. Geologist David Love, as interpreted by John McPhee, "Annals of the Former World," *New Yorker*, February 24, 1986, 67.

2. Homer, *The Odyssey*, trans. Robert Fitzgerald (New York: Doubleday, Anchor edition, 1963), 167; also Gustav Schwab, *Gods and Heroes: Myths and Epics of Ancient Greece* (New York: Pantheon, 1946), 651–52; and Marq de Villiers, *Windswept: The Story of Wind and Weather* (New York: Walker and Company, 2006): 14–15. It is, of course, from Homer's Aeolus that the aeolian science, the study of atmospheric winds, draws its name.

3. Geoffrey Irwin, *The Prehistoric Exploration and Colonization of the Pacific* (Cambridge: Cambridge University Press, 1992), 5, 25–30.

4. Lionel Casson, *Ships and Seamanship in the Ancient World* (Princeton: Princeton University Press, 1971), 18–22; Henry Hodges, *Technology in the Ancient World* (New York: Knopf, 1970): 95–96.

5. Todd Woody, "The Future of Wind Power," *Fortune*, October 27, 2008, 59–66; http://en.wikipedia.org/wiki/SkySails.

6. Edward J. Kealey, *Harvesting the Air: Windmill Pioneers in Twelfth-Century England* (Berkeley: University of California Press, 1987), 2.

7. Ahmad Y. al-Hassan and Donald Hill, *Islamic Technology* (New York: Cambridge University Press, 1986), 55; Charles Singer et. al., eds., *A History of Technology*, 5 vols. (Oxford: Clarendon, 1956), 2:615–16; Michael Harvergon, *Persian Windmills* (Reading, England: International Molinological Society, 1991).

8. Terry S. Reynolds, *Stronger Than a Hundred Men: A History of the Vertical Wheel* (Baltimore: Johns Hopkins University Press, 1983), 48.

9. John Reynolds, *Windmills and Waterwheels* (New York: Praeger, 1975), 89.

10. Kealey, *Harvesting the Air*, 132–53.

11. Jean-Claude Debier, Jean-Paul Deleage, and Daniel Hemery, *In the Servitude of Power: Energy and Civilisation through the Ages*, trans. John Barzman (Atlantic Highlands, N.J.: Zed Books, 1991), 78.

12. Lynn White, Jr., *Medieval Technology and Social Change* (Oxford: Oxford University Press, 1962), 22.

13. Lewis Mumford, *Technics and Civilization* (New York: Harcourt, Brace, 1934), 112–13.

14. T. Lindsay Baker's *A Field Guide to American Windmills* (Norman: University of Oklahoma Press, 1985) is the authoritative book on the American windmill. He presents types, brands, and numbers in exceptional detail.

15. Brian Hayes's wonderful book *Infrastructure: The Book of Everything for the Industrial Landscape* (New York: W. W. Norton, 2005) devotes one page (220) to the American windmill, as he examines every possible industrial use of the American landscape.

16. Quoted in Marc Bloch, *The Historian's Craft* (New York: Knopf, 1959), 66.

17. Andrew Carnegie, "The Conservation of Ores and Related Minerals," *Proceedings of a Conference of Governors, May 13–15, 1908* (Washington: Government Printing Office, 1909), 23–24.

18. Heather Rogers, "Current Thinking," *New York Times Magazine*, June 3, 2007, 18; Peter Asmus, *Reaping the Wind: How Mechanical Wizards, Visionaries, and Profiteers Helped Shape Our Energy Future* (Washington, D. C.: Island Press, 2001), 34.

19. William Thomson, "The Sources of Energy in Nature," *Engineering* (London) 32 (September 23, 1881): 321–22.

20. "Mr. Brush's Windmill Dynamo," *Scientific American* 63 (December 20, 1890): 389.

21. Cleveland *Plain-Dealer*, June 16, 1929.

22. For a more complete story of Brush and his wind dynamo, see Robert W. Righter, *Wind Energy in America: A History* (Norman: University of Oklahoma Press, 1996), 42–58.

23. Clark Spence, "Early Uses of Electricity in American Agriculture," *Technology and Culture* 3 (Spring, 1962): 152. Spence quotes from the editor of the *Rural New-Yorker* 47 (June 9, 1888): 386.

24. F. E. Powell, *Windmills and Wind Motors: How to Build and Run Them* (New York: Spon and Chamberlain, 1918), iii.

25. For more on the Jacobs brothers and their turbine, see Robert W. Righter, "Reaping the Wind: The Jacobs Brothers, Montana's Pioneer 'Windsmiths,'" *Montana: The Magazine of Western History* 46 (Winter, 1996): 38–49.

26. David E. Nye, *Electrifying America: Social Meaning of a New Technology, 1880–1940*, (Cambridge, Mass., MIT Press, 1990), 334.

27. House of Representatives, *Hearing on Rural Electrification Planning before a Subcommittee of the Committee on Interstate and Foreign Commerce*, June 13–15,

October 16–19, 30, November 2, 1945, H.R. 1742, 79th Cong., lst Sess., 107–15. The Wincharger Company, which Weinig represented, manufactured approximately 400,000 wind turbines of all sizes, distributing them worldwide.

28. See Palmer Cosslett Putnam, *Putnam's Power from the Wind* (New York: Van Nostrand Reinhold, 1948). A revised edition, edited by Gerald W. Knoeppl, appeared in 1982. Also see Righter, *Wind Energy in America*, 126–45.

29. Palmer C. Putnam, "Wind-Turbine Power Plant Will Be Rebuilt," *Power* 89 (June, 1945): 68.

30. See Percy H. Thomas, *The Wind Power Aerogenerator, Twin-Wheel Type* (Washington, D.C.: U.S. Federal Power Commission, 1946), 5, exhibit IV; and House of Representatives, Committee on Interior and Insular Affairs, *Hearing of H.R. 4286, A Bill . . . to Determine and Demonstrate the Economic Feasibility of Producing Electric Power and Energy by Means of a Wind-Driven Generator,"* September 19, 1951, 2–5, 27–29.

31. Asmus, *Reaping the Wind*, 47–51: "Wind Power Pioneer Interview: Bill Heronemus, University of Massachusetts," found at www.eere.energy.gov.

32. "Wind Power Pioneer Interview: Bill Heronemus."

33. The Bureau of Reclamation was wildly optimistic with regard to the number of homes to be powered. Using what I consider to be fair home use averages and wind turbine capacity (25%), I would reduce the 67,000 figure to 9,124.

34. Account based on the following articles: *Laramie Daily Boomerang*, May 1, 1985, September 27, 1986; "Nobody Wants 350-foot Windmill Because of 1.5 Million Repair Bill, " *Denver Post*, May 30, 1985; Tody F. Marlatt, "Malfunction Shatters World's Largest Wind Turbine," *Medicine Bow*, Wyoming, January 20, 1994; *Laramie Daily Boomerang*, Sunday, January 16, 1994.

35. State of California, *Statutes and Amendments to the Codes*, 1978, vol. 3, chapter 1159, 3557.

36. Interview with Paul Gipe by the author, November 11, 1992.

37. Advertisement provided to the author by Robert Kahn and Company.

38. *Desert Sun*, January 9, 1988, in wind energy clippings file, Palms Springs Public Library.

Chapter 2. How Have These Large Turbines Evolved?

1. I believe the largest turbine to date is the Enercon E-126, a 6-megawatt turbine constructed in Emden, Germany. The rotor diameter is 417 feet and the swept area is 12,668 square meters, equivalent to 3.13 acres. The total height is 651 feet.

2. It is not easy figuring how wind farm developers compile the number of houses served. I have based some of my figures on material provided by the North Texas Wind Resistance Alliance.

3. Sears Roebuck and Company catalog, Spring–Fall, 1946, 646, 1426. By 1951 the catalogue no longer listed small wind turbines for sale.

4. See *Scientific American* 49 (1883). The magazine was indexed, making articles on wind energy easy to locate.

5. "Mr. Brush's Windmill Dynamo," *Scientific American* 63 (December 20, 1890): 389.

6. For more on Brush and his wind turbine see Righter, *Wind Energy in America*, 42–58.

7. Senate, *Hearings before the Committee on Interior and Insular Affairs Pursuant to S. Res. 45, A National Fuels and Energy Policy Study*, 93rd Cong., 1st Sess., 11. President Nixon's energy message in its entirety is on pages 3 to 20.

8. *New York Times*, June 27, 1976.

9. Richard F. Hirsh, *Power Loss: The Origins of Deregulation and Restructuring in the American Electric Utility System* (Cambridge, Mass.: MIT Press, 1999): 9–10.

10. The one notable exception is the two-bladed Nordic N1000. The Swedish Company has long advocated the two-blade turbine as less expensive to construct and maintain. There were major noise problems with earlier models, but a company representatives said the new 1 MW turbine being introduced in the summer of 2009 has noise levels similar to the three-bladed turbines. *Wind Today* (1st quarter, 2009): 78.

11. Our papers were published in Martin J. Pasqualetti, Paul Gipe, and Robert W. Righter, eds., *Wind Power in View: Energy Landscapes in a Crowded World* (San Diego: Academic Press, 2002). The papers cover both American and European wind energy landscapes, and this is the only book, to my knowledge, that deals in depth with the NIMBY response. Of course I still wonder about the 2.5 and 3.0 MW turbines as they age. Major repair in a nacelle at 250 feet will not be easy and will definitely be expensive.

12. Statistics from www.wind-watch.org.

13. www.gepower.com/prod_serv/products/wind_turbines.

14. See *Wind Today* (4th quarter, 2008): 82–86, for a chart of projects, developers, and turbines being installed. For Gamesa see www.wind-watch.org/news/2009/01/09.

15. Joanna Lake, "Industrial Wind, Corporate Vandalism," *Burlington Free Press*, April 3, 2005, in www.aweo.org/Lake.html; E. F. Schumacher, *Small Is Beautiful: Economics as if People Mattered* (New York: Harper and Row, 1973).

16. Bill Roorbach, "Big Wind: Does a Four-Percent Share of Maine's Power Grid Warrant Turbines atop Kibby Mountain?" *Down East*, in www.wind-watch.org/news/2008/02/12; Lake, "Industrial Wind, Corporate Vandalism."

17. Sandia Laboratory in Albuquerque did considerable experimentation with the "eggbeater" design in the 1980s.

18. Paul Gipe, *Wind Energy Comes of Age* (New York: John Wiley, 1995), 169

19. Righter, *Wind Energy in America*, 274.

20. See Pasqualetti et. al., *Wind Power in View.*

21. See www.Makanipower.com.

22. David Owens, "The Inventor's Dilemma," *New Yorker*, May 17, 2010, 45.

23. *General and Special Laws, Texas*, vol. 3., 76th Legislature, Regular Session, 1999, S.B. 7, "Relating to Electric Utility Restructuring," sec. 39.904, p. 2598.

24. See Thomas Friedman, "Salvage Your Legacy by Ending Our Oil Addiction, Mr. President," *Jackson Hole Daily*, December 16–17, 2006, 5.

25. For Texas wind energy statistics see the American Wind Energy Association website, www.awea.org/projects/texas.html, updated September 30, 2006; Dorothy Scarborough, *The Wind* (New York: Harper, 1925; reprint, New York: Grosset and Dunlap, 1979).

26. "Interview with Rick Perry, Governor of Texas," *Wind Today* 3, no. 2 (2nd quarter, 2008): 23.

27. CNBC interview with Boone Pickens, February 22, 2008; also www.nytimes.com/2008/02/23/business/23wind.

28. "Pecos County Wind Farms: Local Economic Impact," fact sheet published by the town of Fort Stockton's economic development corporation.

29. Interview with Doug May by author, Fort Stockton, Pecos County, Texas, February 15, 2007.

30. Ibid.

31. Interview with Robert and Marjorie Bichsel by author, February 13, 2007.

32. "Interview with Rick Perry, Governor of Texas," *Wind Today* 3, no. 2 (2nd quarter, 2008): 23.

33. Just why the judge ruled out visual pollution is not clear. The ruling may find an explanation in Texas oil history of the days when thousands of oil pump jacks covered the West Texas landscape.

34. Thomas Korosec, "Turbines Generate Opposition," *Houston Chronicle*, February 5, 2007, A-1, 4.

35. Don Graham, *Kings of Texas: The 150-Year Saga of an American Ranching Empire* (Hoboken, N.J.: John Wiley, 2003), 22.

36. *Dallas Morning News*, August 24, 2006.

37. *Dallas Morning News*, September 9, 2006.

38. *Houston Chronicle*, June 3, 2008.

39. Article from *The Monitor* in National Wind Watch, www.wind-watch.org/news/2009/01/08.

40. *Dallas Morning News*, July 6, 2008, Business section.

41. Bill Hanna, "Wind-Power Push Raising Ire," *Fort Worth Star-Telegram*, November 26, 2006.

42. Ibid.; *New York Times*, July 19, 2008, Business section; *Jackson Hole Daily*, July 18, 2008, 23.

43. Williams and Whitcomb, *Cape Wind*.

44. www.capewind.org.

45. Williams and Whitcomb, *Cape Wind*, 302.

46. Abby Goodnough, "Wind Farm off Cape Cod Clears Hurdle," *New York Times*, January 17, 2009.

47. "Six Turbines Doing the Work of Ninety," *Wind Today* 3, no. 2 (2nd quarter, 2008): 36–40.

Chapter 3. The Riddle of Reliability

1. Righter, *Wind Energy in America*, 223–24.

2. Terry S. Reynolds, *Stronger Than a Hundred Men*, 158.

3. Walter Prescott Webb, *The Great Plains* (New York: Grosset and Dunlap, 1971), 337–47. Also see Baker's definitive *Field Guide to American Windmills*.

4. Report in *Alternative Sources of Energy* 50 (July–August 1981): 38.

5. Some years ago I had conversations and correspondence with Matthias Heymann, then a student at the Deutsches Museum, Munich. He provided me with two papers: "Why Were the Danes Best: Wind Turbines in Denmark, West Germany and the USA, 1945–1985," and "Theoretical Knowledge and Experimental Skill: What Made Wind Energy Converter Successful in History and at Present?" I have found his work convincing.

6. See Righter, *Wind Energy in America*, 219–20, 180–82.

7. Heymann, "Why Were the Danes Best," fig. 3, 5.

8. *Desert Sun* (Palm Springs), February 23, 1986; interview by author with Dave Kelly, SeaWest field manager, Tehachapi, November 10, 1992.

9. Paul Gipe, *Wind Power: Renewable Energy for Home, Farm, and Business* (Vermont: Chelsea Green Publishing Company, 2004), 98.

10. www.wind-watch.org/news/2008/03/27.

11. These figures are the result of talks and e-mails with Mark Haller, a wind consultant for twenty-five years, and Paul Gipe, wind energy consultant and prolific author, April 18–22, 2008.

12. See *Cape Wind: Money, Celebrity, Class, Politics, and the Battle for Our Energy Future on Nantucket Sound* (New York: Public Affairs, 2007): 202–204.

13. www.3tiergroup.com/en/html/wind/wind_overview.html. There are a number of "wind assessment" companies. The 2007 third quarter issue of *Wind Today* lists fifteen.

14. *Wind Today* (3rd quarter, 2007): 62.

15. Quoted in Asmus, *Reaping the Wind*, 72.

16. Righter, *Wind Energy in America*, 215.

17. "Workers Turning to Turbine Repair," *Dallas Morning News*, March 8, 2009, 6D.

18. Bonnie Kreps, producer, *Don't Fence Me In*, film (Wilson, Wyo.: Equipoise Fund, 2008).

19. Interview with Doug May by author, Fort Stockton, Pecos County, Texas, February 15, 2007.

20. Phillip F. Schewe, *Grid: A Journey through the Heart of Our Electrified World* (Washington, D. C.: Joseph Henry Press, 2007): 158–59.

21. See chapter 2 for more details.

Chapter 4. Tying into the Grid

1. http://us.f817.mail.yahoo.com/ym/showletter/2/6/2008/.

2. *Dallas Morning News*, April 3, 2008, 6D.

3. Hayes, *Infrastructure*, 229–30.

4. There are essentially eight transmission systems (grids) in the United States. I list them here with their primary geographical areas: Electric Reliability Council of Texas (ERCOT, most of Texas); Florida Reliability Coordinating Council (FRCC, Florida); Midwest Reliability Organization (MRO, northern Midwest and midwestern Canada); Northeast Power Coordinating Council (NPCC, New York, New England, and eastern Canada); Reliability First Corporation (RFC, mainly Michigan, Indiana, Ohio, and Pennsylvania); SERC Reliability Corporation (SERC, southern states except Florida): Southwest Power Pool, Inc. (SPP, mainly Kansas and Oklahoma); and Western Electricity Coordinating Council (WECS, all of the American West states and western Canada). These regional entities are under the general umbrella of the North American Electric Reliability Corporation, based in Princeton, New Jersey. For more information go to www.nerc.com. I am in debt to Warren Lasher, an electrical engineer with ERCOT, for helping me understand many issues related to the grid.

5. The finest description of the Giant Power proposal can be found in Thomas P. Hughes, *Networks of Power: Electrification in Western Society, 1880–1930* (Baltimore: Johns Hopkins University Press, 1983), 296–313.

6. Gifford Pinchot, *The Power Monopoly: Its Make-up and Its Menace* (Milford, Pa.: Privately printed, 1928), 1. Writing at the close of the unsuccessful Giant Power effort, Pinchot was bitter, and this rare book—a copy is in the Huntington Library, San Marino, California—reflects his disappointment and anger.

7. *Report of the Giant Power Survey Board to the General Assembly of the Commonwealth of Pennsylvania* (Harrisburg, Pa.: Telegraph Printing Company, 1925), iv–v, quoted in Hughes, *Networks of Power*, 298.

8. Statistics from Hughes, *Networks of Power*, 307; L. J. Fletcher, "Rural Electric Service from the Western Standpoint," *Agricultural Engineering* 7 (December 1926): 407.

9. Hughes, *Networks of Power*, 305.

10. Ibid., 391–93, 401. Hughes makes a convincing case that these holding companies were not promoted by banks and investors as much as by engineers and company managers, who understood the efficiency of the holding company in managing these huge grid systems.

11. John B. Wilbur, "The Smith-Putnam Wind Turbine Project," *Boston Society of Civil Engineers Journal* 29 (July 1942): 217.

12. John Makansi, *Lights Out: The Electricity Crisis, the Global Economy, and What It Means to You* (New York: John Wiley, 2007), 20–21.

13. U.S. Code, Congressional and Administrative News, 95th Cong., 2nd Sess., 1978, *Laws*, vol. 2, Public Law 95-617, sec. 210.

14. See PURPA *Handbook for Independent Electric Power Producers* (Washington, D.C.: American Wind Energy Association, 1992). Anyone considering a putting up a small turbine should read Paul Gipe's discussion of PURPA in *Wind Power*, 205–10.

15. Righter, *Wind Energy in America*, 185–88.

16. I depend a great deal on Hayes, *Infrastructure*, 229–75. He does a remarkable job of making incomprehensible concepts comprehensible.

17. Ibid., 260.

18. Interview by author with Ken Starcher, West Texas State University, February 13, 2007.

19. www.wind-watch.org/news/2008/05/17/wind-project-could-be-expanding/print/.

20. Judith Lewis, "Walking on a Wire," *High Country News* 40, no. 11 (June 9, 2008): 10–13, 27.

21. Ibid., 12.

22. As compared with at least ten years for a coal-fired or nuclear plant.

23. Friedman, *Hot, Flat, and Crowded*, 192–93.

24. "The Big Energy Gamble: California's Efforts to Ensure a Sustainable Energy Future," *NOVA*, Public Broadcasting Service, January 20, 2009.

25. Figures from http://en.wikipedia.org/wiki/InterstateHighwaySystem.

26. "The Next President's First Task—A Manifesto," *Vanity Fair*, May, 2008, 228–29. Robert Kennedy, Jr., and Ted Kennedy have not been friends to wind energy. Their opposition to the Cape Wind project is discussed at length in chapter 7.

27. "Wind: The Power, the Promise, the Business," *Businessweek*, July 2, 2008, 47.

28. Makansi, *Lights Out*, 64.

29. Ibid., 264.

30. For the Electricity Storage Association, see www.electricitystorage.org.

31. See www.isepa.com.

32. Public Utility Law Project, www.pulp.tc/html/_smart_electric_meters_debate .html.

33. http://en. wikipedia.org/wiki/Smart_meter.

34. For more discussion see Makansi, *Lights Out*, 261–62.

Chapter 5. Can the Government Help?

1. "Wind Power: The Wave of the Future?" *Justice Talking*, National Public Radio, Cape Cod, Mass. Summer 2005.

2. Glenn Schleede, "'Big Money' Discovers the Huge Tax Breaks and Subsidies for Wind Energy While Taxpayers and Electric Customers Pick up the Tab," April 14, 2005, www.aweo.org, a website dedicated to opposing wind energy. The site fails to define AWEO, but I assume it means American Wind Energy Organization.

3. *Wind Energy Weekly* 27, no. 1325 (February 13, 2009): 2; AWEA Legislative Alert (e-mail), January 12, 2009.

4. *Wind Today* 4, no. 1 (1st quarter, 2009): 54.

5. "Congress Seen Backing Renewable Energy Standard," www.wind-watch.org/news/2009/02/10.

6. *Dallas Morning News*, April 2, 2008, 3D.

7. Vijay V. Vaitheeswaran, *Power to the People: How the Coming Energy Revolution Will Transform an Industry, Change Our Lives, and Maybe Even Save the Planet* (New York: Farrar, Straus and Giroux, 2003), 199–200.

8. "Wind Power: The Wave of the Future," *Justice Talking*, National Public Radio, Cape Cod, Mass., Summer 2005.

9. Righter, *Wind Energy in America*, 150.

10. *Federal Energy Subsidies: Not All Technologies Are Equal*, report released by Renewable Energy Policy Project (REPP), July 2000. See www.repp.org.

11. Carl Levesque, "What Is the Percentage of Federal Subsidies Allotted for Wind Power?" in www.renewableenergyaccess.com/rea/news/story?id=48070.

12. www.nrel.gov/wind/facilities.html.

13. *Wind Today* 4, no. 1 (1st quarter, 2009): 8.

14. www.stpns.net/view/article.

15. "Oregon Is Exceptionally Generous with Green-Energy Subsides," *Oregonian* (Portland), January 2, 2009.

16. See State-Level Renewable Energy Portfolio Standards chart, in www.AWEA.org

17. *Windletter* 27, no. 12 (December 2008): 12–13.

18. "Stop Hiding the Terms of Wind Leases," www.wind-watch.org/news/2009/04/22.

19. Disclosure requirements and the terms of leases are rapidly changing to openness. For instance, the Ruckelshaus Institute of Environment and Natural Resources, associated with the University of Wyoming, has published a guide titled "*Commercial Wind Energy Development in Wyoming: A Guide for Landowners*" (May 2009).

20. *Desert Sun* (Palm Springs), June 3, 1989, July 1, 1989; also Ros Davidson, "Sonny Bono Takes His War on Windmills to Washington," *Windpower Monthly* 5 (July 1989): 21.

21. Statistics from "The Difference Wind Makes," www.awea.com.

22. Jeff Anthony, "Renewable Rewards," *Windletter* 27, no. 12 (December 2008): 1–5.

23. www.wind-watch.org/news/2009/01/09.

24. See Gipe, *Wind Energy Comes of Age*, 38, and Gipe, *Wind Power*, 210–16.

25. Gipe, *Wind Power*, 211.

26. "Wind Power Advocate Interview: Lisa Daniels, Windustry," www.windpowerinamerica.gov.

27. Gipe, *Wind Power*, 58–60.

28. Bryan Walsh-Samso, "The Gusty Superpower: How Denmark's Green Energy Initiatives Power Its Economy," *Time*, March 16, 2009.

29. Martin Hoppe-Kilpper and Urta Steinhauser, "Wind Landscapes in the German Milieu," in *Wind Power in View*, ed. Martin J. Pasqualetti, Paul Gipe, and Robert W. Righter (San Diego: Academic Press, 2002), 84–85.

30. Senate, *Hearings before the Committee on Interior and Insular Affairs Pursuant to S. Res. 45, A National Fuels and Energy Policy Study*, 93rd Cong., lst Sess. President Nixon's energy message, in its entirety, is on pages 3 to 20.

31. *PURPA Handbook for Independent Electric Power Producers*, 5–7.

32. Ibid., 6–7. See Hirsh, *Power Loss*, 73–100, for an enlightening discussion of the passage and the importance of PURPA.

Chapter 6. "Not in My Backyard"

1. Sylvia White, "Towers Multiply, and Environment Is Gone to the Wind," *Los Angeles Times*, November 26, 1984, sec. II, 5.

2. Seth Zuckerman, "Winds of Change," *Image*, September 20, 1987, 30.

3. www.wind-watch.org.

4. Pasqualetti et al., *Wind Power in View*.

5. Christoph Schwahn, "Landscape and Policy in the North Sea Marshes [of Germany]," in Pasqualetti et al., *Wind Power in View*, 136–40.

6. Gordon Brittan, "A View from Lake Como," in Pasqualetti et al., *Wind Power in View*, 215–16.

7. *Burlington Free Press* (Vermont), April 3, 2005, G. She was referring to E. F. Schumacher's classic *Small Is Beautiful*.

8. Laurence Short, "Wind Power and English Landscape Identity," in Pasqualetti et al., *Wind Power in View*, 43.

9. Karen Hammarlund, "Society and Wind Power in Sweden," in Pasqualetti et al., *Wind Power in View*, 107.

10. See Gipe, *Wind Energy Comes of Age*, 446–50; also Righter, *Wind Energy in America*, 236–39.

11. Alston Chase, "California Wind Farm Fight Generates Unusual Foes," *Jackson Hole Daily* (August 23, 1989).

12. www.wind-watch.org/documents/bird-fatality-study-at-altamont-pass-wind-resource.

13. "Wind Energy and Wildlife: Frequently Asked Questions," www.awea.org.

14. Ibid.

15. www.wind-watch.org/news/2008/04/24/locations-of-wind-turbines-has-some-birders.

16. Ibid. Fitzpatrick's suggestion that the new larger turbines act like "ceiling fans" neglects the fact that the tip speed is usually between 120 and 160 miles per hour.

17. *The Monitor*, January 8, 2009, www.wind-watch.org/news/2009/01/08.

18. "Audubon Society Requests Study on Danger to Birds," www.windwatch.org/news/ 2008/01/28.

19. See National Research Council of the National Academies, *Environmental Impacts of Wind-Energy Projects* (Washington, D.C.: National Academies Press, 2007), 122–39.

20. "Rare Birds Could Be Threatened by Growth of Wind Farms," www.wind-watch.org/news/2008/02/27.

21. Ibid.

22. Robert L. Thayer, Jr., "The Aesthetics of Wind Energy in the United States: Case Studies in Public Perception," *Proceedings*, European Community Wind Energy Conference, Herning Congress Centre, Denmark, June 6–10, 1988, 167–68, 473.

23. "Turbine Noise Irks Some in Ubly Area: Residents Say Windmills Are Louder Than They Thought They'd Be," www.wind-watch.org/news/2009/02/11.

24. "Myth No. 4 Is No Myth; Wind Turbines Are Noisy," www.wind-watch.org/news/2008/02/05.

25. Ibid.

26. "Wind: The Power, the Promise, the Business," *Businessweek*, July 7, 2008, 47.

27. "Adirondack Group Opposes Windmills," www.wind-watch.org/news/2008/02/20.

28. Commentary by Bill Mckibbon, "Wind Power: The Wave of the Future," *Justice Talking*, National Public Radio, September 8, 2006.

Chapter 7. Addressing NIMBY

1. Yi-Fu Tuan, *Topophilia: A Study of Environmental Perception, Attitudes, and Values* (New York: Columbia University Press, [1976] 1990 ed.), 45–58.

2. Hayes, *Infrastructure*.

3. Robert Thayer, Jr., and Carla M. Freeman, "Altamont: Public Perceptions of a Wind Energy Landscape," *Landscape and Urban Planning* 14 (1987): 393.

4. Robert Thayer, *Gray World, Green Heart* (New York: John Wiley, 1994), 55.

5. Certainly this "guilt" was partly responsible for the Wilderness Act of 1964, an effort to save a portion of America from human imprint.

6. See *Commercial Wind Energy Development in Wyoming: A Guide for Landowners* (Laramie, Wyo.: Ruckelshaus Institute of Environment and Natural Resources, University of Wyoming, 2009). This booklet contains much information for a landowner who is considering leasing land for turbines. It also contains a wind energy lease agreement template.

7. Karin Hammarlund, "Society and Wind Power in Sweden," in Pasqualetti et al., *Windpower in View*, 106–108.

8. Williams and Whitcomb, *Cape Wind*. Ironically, at this writing, Robert Kennedy, CEO of the Boston Citizens Utility Group, is negotiating with the Navajo Nation to build a 500 MW wind farm on the reservation.

9. Martin J. Pasqualetti, "Living with Wind Power in a Hostile Landscape," in Pasqualetti et al., *Wind Power in View*, 168.

10. Letter from Sue Sliwinski of Sardinia, New York, dated September 27, 2005, www.aweo.org/Sliwinski.html. I assume that severe "stray voltage" refers to static electricity.

11. www.wind-watch.org/news/2008/04/11/resident-talks-about-living-near-turbines/print/.

12. www.wind-watch.org/news/2008/04/11/wind-energy-hearing-attracts-critics/print/.

13. www.mapleridgewind.com.

14. David Baron, "Wind Farm Buffets Family, Town Relations," National Public Radio, April 22, 2008, www.npr.org.

15. Ibid.

16. "Wind Energy Sweeps Plains: Course Correction in Order," *New Mexican* (Santa Fe), January 11, 2009; Staci Matlock, "Wind Chill: Rural Residents Worry about Impact of Lightly Regulated Industry," *New Mexican*, January 11, 2009.

17. Paul Gipe, "Landowner Leases and Royalty Payments," www.wind-works.org.

18. Ibid.

19. Paul Gipe, "Wind Turbine Envy and Land Leasing Pooling," www.wind-works.org.

20. See Florence Williams, "Plains Sense: Frank and Deborah Popper's 'Buffalo Commons' is Creeping toward Reality," *High Country News*, January 15, 2001, and J. Michael Hayden, "Were the Poppers Right?" *Online Journal of Rural Research and Policy*, issue 2, June 1, 2007.

21. "Winding Up to Blows," www.wind-watch.org/news/2008/02/21.

Chapter 8. Small Turbines and Appropriate Technology

1. "Small Turbines for Rural Development," www.bergey.com. I rely on much of the information from Bergey Windpower, located in Norman, Oklahoma. The company has made small wind turbines since the 1970s. Also, it has a global reach, with representatives in more than fifty countries.

2. "AWEA Small Wind Turbine Global Market Study, 2008," www.awea.org.

3. "New Metering and Related Utility Issues," www.bergey.com.

4. See Martin Merzer, "Windmill Power Up in Air," *Denver Post*, June 15, 1977.

5. *Buying a Small Wind Electric System: A California Consumer's Guide*, (Sacramento; California Energy Program, 2002), 9.

6. State of California, *Statutes and Amendments to the Codes*, 1978, vol. 3, chapter 1159, 3557.

7. Given the fiscal crisis in California, one wonders if this rebate system will survive.

8. *Buying a Small Wind Electric System*, 8.

9. "Small Wind Systems in California: Let the Wind Pay Your Electric Bill," www.bergey.com.

10. The Southwest Windpower site may be found at www.windenergy.com.

11. F. C. Fenton and D. E. Wiant, "Rural Electrification Surveys of Harvey and Dickinson Counties," *Kansas State College Bulletin* 24 (May 1, 1940), 47 pp.

12. Interview with Joe and Mary Spinhirne by author, November 6, 1985.

13. See Barry M. Casper and Paul David Wellstone, *Powerline: The First Battle of America's Energy War* (Amherst: University of Massachusetts Press, 1981), for the remarkable story of a group of committed farmers who fought, in both word and deed, the REA bureaucracy's dedication to huge central policy schemes.

14. Michael Bergey, "A Primer on Small Turbines," www.bergey.com.

15. Mick Sagrillo, "A Call for a Sensible Approach to Small Wind," *Windletter* 27, no. 12 (December 2008): 6–10.

16. In 2008 twenty-five U.S. manufacturers responded to the AWEA survey. "AWEA Small Wind Turbine Global Market Study, 2008," www.awea.org.

17. There are questions regarding the financial stability of the company. Anyone interested should look closely at the most recent information. See websites for Helix Wind (www.helixwind.com/en/) and Windside Oy (www.windside.com); also Todd Woody, "The Future of Wind Power," *Fortune*, October 27, 2008, 59–65.

18. Mick Sagrillo, "Urban Turbines," *Windletter* 28, no. 2 (February 2009): 6–7.

19. Sagrillo, "A Call for a Sensible Approach," 9–10.

20. See "Backyard Windmill Creates Buzz in Small Town," *Dallas Morning News*, July 26, 2008, 1L.

21. Gipe, *Wind Power*, 199.

22. Katie Arnold, "Green Machines," noted in a Western regional real estate magazine.

23. William H. Kemp, "An Off-Grid Primer," 2003, www.bergey.com/School/Kemp.Article.pdf.

24. "Small Turbines for Rural Development," www.bergey.com.

25. "Afghanistan's First Windfarm Uses Bergey Turbines," www.bergey.com.

Chapter 9. Other Solutions

1. See Terry S. Reynolds, *Stronger Than a Hundred Men*.

2. Ibid., 48.

3. Harold H. Schobert, *Energy and Society* (New York: Taylor and Francis, 2002), 199.

4. See J. S. Holliday, *Rush for Riches: Gold Fever and the Making of California* (Berkeley: University of California Press, 1999), 274–75; Malcomb Rohrbough, *Days of Gold: The California Gold Rush and the American Nation* (Berkeley: University of California Press, 1997), 203; Robert W. Righter, *The Battle over Hetch Hetchy: America's Most Controversial Dam and the Birth of Modern Environmentalism* (New York: Oxford University Press, 2005), 31–33.

5. Schobert, *Energy and Society*, 198.

6. www.ucsusa.org/clean_energy/coalvswind/index.html.

7. See U.S. Energy Information Administration, Electric Power Annual, 2008, www.eia.doe.gov/eneaf/electricity/epa/epa.sum.html; also see http://en.wikipedia.org/wiki/Clean_coal.

8. www.fossil.energy.gov/programs/powersystems/cleancoal.

9. The American Wind Energy Association claims that $35 billion have been paid out over the last thirty years; www.awea.org/faq/wwt_costs.html.

10. See www.eia.doe.gov/fuelelectric.html.

11. www.pickensplan.com.

12. Twenty-seven cents versus 15 cents. See "AWEA Small Wind Turbine Global Market Study, 2008," 15–16.

13. Debier et al., In the Servitude of Power, 170, 186.

14. Richard Munson, The Power Makers (Emmaus, Pa.: Rodale, 1985), 105.

15. http://en.wikipedia.org/wiki/Nuclear power.

16. Statistics from the Geothermal Energy Association, www.geo-energy.org.

17. Ibid.

18. Ibid.

19. Vaitheeswaran, Power to the People, 224–31, 14.

20. Ibid., 227–31.

21. The story of PG&E's change of heart with regard to nuclear and alternative energy is well told by David Roe, Dynamos and Virgins (New York: Random House, 1984); also see John Wills, Conservation Fallout: Nuclear Protest at Diablo Canyon (Reno: University of Nevada Press, 2006).

22. "The Source: Utility Enterprise Management," www.ae2s.com/pdf/source.

23. For more information on smart meters, see http://en.wikipedia.org/wiki/Smart meter.

24. See Roe, Dynamos and Virgins, 71, 120.

25. Ben Elgin and Diana Holden, "Green Power: Buyers Beware," Businessweek, September 29, 2008, 68–70.

Chapter 10. Living with the Wind and the Turbines

1. Scarborough, The Wind, 337.

2. Quoted in De Villiers, Windswept, 22.

3. www.cloudwall.com.

4. De Villiers, Windswept, 3.

5. Owen Wister to his mother, July 28, 1885, in Francis K. W. Stokes, My Father, Owen Wister (Laramie, Wyo.: n.p., 1952), 40.

6. Walter Prescott Webb in The Great Plains attributed the settlement of land west of the hundredth meridian to three technologies: the six-gun, barbed wire, and the windmill—the technologies necessary for successful cattle ranches.

7. Tuan, Topophilia, 108–109.

8. Robert W. Righter, "Ekoskeletal Outer-Space Creations," in Pasqualetti et al., Wind Power in View, 20–22.

9. Robert Thayer and Heather Hansen, "Consumer Attitudes and Choice in Local Energy Development," Department of Environmental Design, University of California, Davis, May 1989, 17–19. Also see Thayer, Jr. *Gray World, Green Heart*, 74–75, and Maarten Wolsink, "Attitudes and Expectations about Wind Turbines and Wind Farms," *Wind Engineering* 13, no. 4 (1989): 196–206.

10. National Research Council, *Environmental Impacts of Wind-Energy Projects*, 140–41.

11. Ibid., 147.

12. "Wind Turbines Help Create Energy but Also Disturb People's Homes," www .wind-watch.org/news/2009/04/12.

13. Wind turbine noise measurements are beyond the scope of this work. However, they are extremely important in determining setbacks for turbines. One must consider the vibrations moving through the air, where factors such as atmospheric conditions and air density play a role. Thorough testing is necessary; most important, it must be performed *before* the turbines are erected. And the testing should be performed by an independent audiologist, not one contracted by the wind farm developer. One noise statistic is easily obtained. Every turbine has moving parts (gears, blades, etc.), and the noise emitting from a turbine can be measured at different speeds. Manufacturers should make this information available not only to wind developers but also to the general public. For a more technical discussion of noise, see National Research Council, *Environmental Impacts of Wind-Energy Projects*, 156–60.

14. "Introduction: Wind Turbine Syndrome," www.windturbinesyndrome.com, posted July 26, 2008.

15. Nina Pierpont, *Wind Energy Syndrome: A Report on a National Experiment* (Lowell, Mass.: King Printing, 2009). It is available through the author or Amazon.

16. "Wind Turbines and Health Problems in Maine," www.windconcernsontario .wordpress.com/2009/03/31.

17. "In a Spin . . . Professor Claims Wind Farm Can Cause Seizures," www.wind -watch.org/news/2008/05/01; "Did Wind Turbines at Whitemoor Contribute to Suicide?" www.wind-watch.org/news/2008/01/26.

18. Harvard Project on American Indian Economic Development, *The State of the Native Nations* (New York: Oxford University Press, 2008), 6, 161–62.

19. Barbara Rose Johnston, Susan Dawson, and Gary Madsin, "Uranium Mining and Milling: Navajo Experiences in the American Southwest," in *Indians and Energy: Exploitation and Opportunity in the American Southwest*, ed. Sherry L. Smith and Brian Frehner (in press), MS 157.

20. Colleen O'Neill, "Jobs and Sovereignty: Tribal Employment Rights and Energy Development in the 20th Century," in Smith and Frehner, *Indians and Energy*, MS 184.

21. Donald L. Fixico, "Understanding the Earth and the Demand on Energy Tribes," in Smith and Frehner, *Indians and Energy*, MS 25.

22. See executive summary, "Energy and Economic Alternatives to the Desert Rock Energy Project," ECCS Consulting for the Din. CARE, January 12, 2008.

23. Felicia Fonseca, "Joseph Kennedy Pushes Wind Energy for Navajo Nation," *Albuquerque Journal*, online edition, April 24, 2008.

24. "Navajo Nation to Develop 500 MW of Wind Power," March 27, 2008, in www .reuters.com/environmental_news.

25. "Energy Justice in Native America and the Next Administration," Policy Statement, Presidential Transition 2009, December 17, 2008, by Honor the Earth, Intertribal Council on Utility Policy, Indigenous Environmental Network, International Indian Treaty Council. Contact: info@honorearth.org.

26. "Native American Interview: Robert Gough, Intertribal Council on Utility Policy (COUP)" December 1, 2004, in www.windpoweringamerica.gov. The tribes include the Cheyenne River; Flandreau Santee; Lower Brule: Mandan, Hidatsa, and Arikara; Omaha; Rosebud; Sisseton; Spirit Lake; and Standing Rock Sioux.

27. Quoted in Felicity Barringer, "Indian Tribes See Profit in Harnessing the Wind for Power," *New York Times*, October 10, 2008.

28. Ibid.

29. See www.NREL.gov/wind.

30. Carl Levesque, "The Little Program with Big Benefits," *Windletter* 28, no. 1 (January, 2009): 1, 4–5.

Conclusion

1. Connie Woodhouse, "Droughts of the Past, Implications for the Future?" in Sherry L. Smith, ed., *The Future of the Southern Plains* (Norman: University of Oklahoma Press, 2003), 106.

2. Williams and Whitcomb, *Cape Wind*, 199–202.

3. www.3tiergroup.com.

4. Interview with Doug May by author, Fort Stockton, Pecos County, Texas, February 15, 2007.

5. Elliott West, "Trails and Footprints: The Past of the Future Southern Plains," in *The Future of the Southern Plains*, ed. Sherry L. Smith (Norman: University of Oklahoma Press, 2003), 40.

6. Hayes, *Infrastructure*, 501–502.

7. *Wind Energy Weekly* 27, no. 1289 (May 9, 2008).

8. *Wind Energy Weekly* 27, no. 1280 (March 7, 2008).

9. "Factcheck: Hot Air on Wind Energy," www.wind-watch.org/news/2009/04/11.

10. "Wind Energy Progress," *Wind Today* 3, no. 4 (4th quarter, 2008): 82–86.

11. "U.S. Wind Energy Installations Top 20,000 MW," www.awea.com.

12. Ibid.

13. "Windpower 2009 Conference and Exhibition," www.awea.com.

14. See Ian L. McHarg, *Design with Nature* (New York: Doubleday, 1971).

15. *Dallas Morning News*, May 15, 2008, 1D.

16. *Wind Power Trail*, a map brochure with windmill and wind turbine facts and site descriptions and locations, funded by Xcel Energy and a number of town chambers of commerce.

17. For the speech see www.repoweramerica.org.

18. Friedman, *Hot, Flat, and Crowded*, 397.

19. Nickolas Confessore, "In Rural New York, Windmills Can Bring Whiff of Corruption," *New York Times*, August 18, 2008.

20. Martin J. Pasqualetti, "Morality, Space, and the Power of Wind-Energy Landscapes," *Geographical Review* 90, no. 3 (July 2000): 392.

21. As quoted in Mary Jo Murphy, "Becoming the Big New Idea: First, Look the Part," *New York Times*, August 24, 2008, sec. 4.

22. Hayes, *Infrastructure*, 3.

23. Murphy, "Becoming the Big New Idea."

24. Friedman, *Hot, Flat, and Crowded*, 205.

25. Ibid., 230.

Suggested Readings

Asmus, Peter. *Reaping the Wind: How Mechanical Wizards, Visionaries, and Profiteers Helped Shape Our Energy Future*. Washington, D.C.: Island Press, 2000.

Baker, T. Lindsay. *A Field Guide to American Windmills*. Norman: University of Oklahoma Press, 1985.

Gipe, Paul. *Wind Energy Comes of Age*. New York: John Wiley, 1995.

——. *Wind Power: Renewable Energy for Home, Farm, and Business*. Rev. ed. White River Junction, Vt.: Chelsea Green, 2004.

Golding, E. W. *The Generation of Electricity by Wind Power*. London: E. & F. N. Spon, 1955.

Pasqualetti, M. J., P. Gipe, and R. W. Righter, eds. *Wind Power in View: Energy Landscapes in a Crowded World*. San Diego: Academic Press, 2002.

Putnam, P. C. *Power from the Wind*. New York: Van Nostrand Reinhold, 1948.

Righter, Robert W. *Wind Power in America: A History*. Norman: University of Oklahoma Press, 1996.

Index

References to illustrations appear in italics.

CPSIA information can be obtained at www.ICGtesting.com
Printed in the USA
243733LV00005B/1/P